T0134577

Springer Theses

Recognizing Outstanding Ph.D. Research

Aims and Scope

The series "Springer Theses" brings together a selection of the very best Ph.D. theses from around the world and across the physical sciences. Nominated and endorsed by two recognized specialists, each published volume has been selected for its scientific excellence and the high impact of its contents for the pertinent field of research. For greater accessibility to non-specialists, the published versions include an extended introduction, as well as a foreword by the student's supervisor explaining the special relevance of the work for the field. As a whole, the series will provide a valuable resource both for newcomers to the research fields described, and for other scientists seeking detailed background information on special questions. Finally, it provides an accredited documentation of the valuable contributions made by today's younger generation of scientists.

Theses are accepted into the series by invited nomination only and must fulfill all of the following criteria

- They must be written in good English.
- The topic should fall within the confines of Chemistry, Physics, Earth Sciences, Engineering and related interdisciplinary fields such as Materials, Nanoscience, Chemical Engineering, Complex Systems and Biophysics.
- The work reported in the thesis must represent a significant scientific advance.
- If the thesis includes previously published material, permission to reproduce this must be gained from the respective copyright holder.
- They must have been examined and passed during the 12 months prior to nomination.
- Each thesis should include a foreword by the supervisor outlining the significance of its content.
- The theses should have a clearly defined structure including an introduction accessible to scientists not expert in that particular field.

More information about this series at http://www.springer.com/series/8790

Alexander Guttridge

Photoassociation of Ultracold CsYb Molecules and Determination of Interspecies Scattering Lengths

Doctoral Thesis accepted by
the Durham University, Durham, UK

Author
Dr. Alexander Guttridge
Department of Physics
Durham University
Durham, UK

Supervisor
Prof. Simon L. Cornish
Department of Physics
Durham University
Durham, UK

ISSN 2190-5053   ISSN 2190-5061   (electronic)
Springer Theses
ISBN 978-3-030-21203-2   ISBN 978-3-030-21201-8   (eBook)
https://doi.org/10.1007/978-3-030-21201-8

This Springer imprint is published by the registered company Springer Nature Switzerland AG
The registered company address is: Gewerbestrasse 11, 6330 Cham, Switzerland

Supervisor's Foreword

The development of ultracold *atomic* gases cooled to temperatures below 1 μK has enabled the study of a staggering range of fascinating phenomena in both few-body and many-body quantum physics. Encouraged by these successes, many researchers seek to bring *molecules* into the ultracold regime, motivated by the rich and diverse properties of molecules. These properties include a complex internal structure associated with vibration and rotation, as well as long-range dipole-dipole interactions stemming from the electric dipole moment of the molecule. As such, ultracold molecules offer a broad range of new and exciting applications ranging from precision measurement of fundamental physics, through ultracold chemistry to quantum simulation of many-body spin systems.

The complex internal structure of molecules presents a considerable challenge for direct cooling techniques. Nevertheless, there has been notable recent success in the laser cooling of a certain class of molecules whose structure leads to limited branching between vibrational levels. However, this approach has not, as yet, produced the density of molecules needed for many applications. Thankfully, another approach exists, whereby pre-cooled atoms are bonded together using subtle association techniques to form molecules deep in the ultracold regime and at high densities where interactions and collisions are important. To date, experiments have been limited to molecules composed of two different alkali-metal atoms.

This thesis lays the groundwork for producing a new class of ultracold molecule by associating an alkali-metal atom and a closed-shell alkaline-earth-like atom, specifically Cs and Yb. This class of molecules exhibit both a magnetic dipole moment and an electric dipole moment in their ground state. This extra degree of freedom opens up new avenues of research including the study of exotic states of matter, the shielding of molecular collisions, the simulation of lattice spin models and tests of fundamental physics.

In detail, the thesis reports the first and only ultracold mixture of Cs and Yb in the world, giving details of the methods used to cool such contrasting atomic species together. Prior to this work, the nature of the interspecies collisions between Cs and Yb atoms were unknown. However, as reported, using sensitive two-colour photoassociation measurements to measure the binding energies of the near-threshold CsYb molecular levels in the electronic ground state has allowed the scattering lengths to be accurately determined for all the Cs–Yb isotopic combinations. As part of this work, the one-photon photoassociation of ultracold Cs^*Yb is also studied, yielding useful information on the excited-state potential. Knowledge of the scattering lengths enables a strategy to be devised to cool both species to quantum degeneracy and, crucially, determines the magnetic field values of the interspecies Feshbach resonances needed for the efficient association of ground-state CsYb molecules. With these results, the prospect of bringing a new molecule into the ultracold regime has become considerably closer.

Durham, UK   Prof. Simon L. Cornish
April 2019

Abstract

This thesis reports the first measurements of the ground-state binding energies of CsYb molecules and the scattering lengths of the Cs+Yb system. The knowledge gained from these measurements will be essential for devising the most efficient route for the creation of rovibrational ground-state CsYb molecules. CsYb molecules in the rovibrational ground state possess both electric and magnetic dipole moments which opens up a wealth of applications in many areas of physics and chemistry.

In addition, we present the setup of a crossed beam optical dipole trap and the investigation of precooling and loading of Yb into the dipole trap. Evaporative cooling in the dipole trap results in the reliable production of Bose-Einstein condensates with 4×10^5 ^{174}Yb atoms. We also describe the necessary changes required to cool fermionic ^{173}Yb atoms and report the production of a six-component degenerate Fermi gas of 8×10^4 ^{173}Yb atoms with a temperature of $0.3\,T_F$.

As well as the ability to cool Yb to degeneracy, we present the production of Bose-Einstein condensates containing 5×10^4 ^{133}Cs atoms. Effective cooling of Cs is achieved using degenerate Raman sideband cooling, which enables 6×10^7 Cs atoms to be cooled to below $2\,\mu$K and polarised in the $|F = 3, m_F = +3\rangle$ state with 90% efficiency.

Finally, we report the production of ultracold heteronuclear Cs*Yb and CsYb molecules using one-photon and two-photon photoassociation respectively. For the electronically excited Cs*Yb molecules we use trap-loss spectroscopy to detect molecular states below the $Cs(^2P_{1/2}) + Yb(^1S_0)$ asymptote. For ^{133}Cs^{174}Yb, we observe 13 rovibrational states with binding energies up to ~ 500 GHz. In addition, we produce ultracold fermionic ^{133}Cs^{173}Yb and bosonic ^{133}Cs^{172}Yb and ^{133}Cs^{170}Yb molecules. From mass scaling, we determine the number of vibrational levels supported by the 2(1/2) excited-state potential to be 154 or 155.

Publications Related to this Thesis

The following publications have resulted from the work presented within this thesis:

- **A versatile dual-species Zeeman slower for caesium and ytterbium**
  S. A. Hopkins, K. L. Butler, A. Guttridge, R. Freytag, S. L. Kemp, E. A. Hinds, M. R. Tarbutt, and S. L. Cornish
  Rev. Sci. Instrum. **87** *043109 (2016)*

- **Direct loading of a large Yb MOT on the $^1S_0 \to {}^3P_1$ transition**
  A. Guttridge, S. A. Hopkins, S. L. Kemp. D. Boddy, R. Freytag, M. P. A. Jones, E. A. Hinds, M. R. Tarbutt, and S. L. Cornish
  J. Phys. B: At. Mol. Opt Phys. **49** *145006 (2016)*

- **Interspecies thermalization in an ultracold mixture of Cs and Yb in an optical trap**
  A. Guttridge, S. A. Hopkins, S. L. Kemp, M. D. Frye, J. M. Hutson, and S. L. Cornish
  Phys. Rev. A **96** *012704 (2017)*

- **Production of ultracold Cs*Yb molecules by photoassociation**
  A. Guttridge, S. A. Hopkins, M. D. Frye, J. J. McFerran, J. M. Hutson, and S. L. Cornish
  Phys. Rev. A **97** *063414 (2018)*

- **Two-photon photoassociation spectroscopy of CsYb: Ground-state interaction potential and interspecies scattering lengths**
  A. Guttridge, M. D. Frye, B. C. Yang, J. M. Hutson, and S. L. Cornish
  Phys. Rev. A **98** *022707 (2018)*

Acknowledgements

Throughout my 4-year journey as a Ph.D. student there have been many people who have supported me and made my time in Durham so enjoyable. There are far too many people to name individually but I thank each and every one of you for your support.

First and foremost I must thank my supervisor Simon Cornish for his endless support and guidance. Every time I have trudged into Simon's office due to something not working in the lab, Simon has always been able to remotivate me and send me off armed with some fresh ideas to tackle the problem. I attribute a large amount of my growth as a physicist to being a student under his supervision and I am hopeful that some of his many great qualities, like his attention to detail and his endless enthusiasm have rubbed off on me a bit.

Second, I must thank the UK Engineering and Physical Sciences Research Council for funding my Ph.D. studentship.

I must also thank the past and present members of team CsYb. Stefan Kemp was a great first mentor and I am very grateful for his patience while showing me the ropes. In addition, I must thank Steve Hopkins for his enormous contribution to the CsYb experiment. Steve's immense technical knowledge, love of hard problems and fantastic grasp of physics were instrumental in the success of the project and also in my own progression as a researcher. I am very thankful that I had such a fantastic mentor and I sincerely hope you enjoy your well-deserved retirement.

John McFerran played a critical role in the progress of the experiment during his 3-month visit and his cheerful Aussie personality was incredibly uplifting throughout the long search for a one-photon photoassociation line. During the writing of this thesis, Jack Segal and Kali Wilson had the unenviable experience of being thrown straight in at the deep end, tasked with operating a complex experiment after less than a week of on-the-job training. Both Jack and Kali are very hard-working and have performed admirably in my stead. I am certain they will make great strides towards producing ground-state CsYb molecules.

I would also like to thank the (now many) past and present members of the Cornish group. Especially Ana Rakonjac, Phil Gregory, Liz Bridge and Dani

Pizzey, all of whom provided a lot of help and guidance throughout my Ph.D. Of course, I must also thank the rest of the current Cornish group: Oliver, Jake, Lewis, Mew, Rahul, Prosenjit, Sarah and Vincent for teaching me a thing or two about baking.

My thanks go to the many members of the wider AtMol group for making it such a friendly work environment. Ever since I started in the group as a summer student 5 years ago, I was certain that this is the place I wanted to do research. Special thanks go to the regular members of the Friday night pub crew for their excellent company and for always keeping me well hydrated.

At this point, I must also thank my secondary supervisor Ifan Hughes. I have benefited greatly from his guidance and his expertise on all things error related. I have enjoyed the many pearls of wisdom that he has shared during our Friday 'supervisory' meetings at the Swan.

I am very thankful to all the technical staff in the Durham physics department for the fantastic work they have done. I am especially grateful to Steve Lishman for sorting out the many rush jobs and urgent repairs I have dumped on his desk.

Collaboration with theory has been essential from the very outset of the CsYb project. Jeremy Hutson and his group have provided guidance as well as a lot of the quantum scattering calculations presented in this thesis. I must thank both Jeremy and Matthew Frye for teaching me many things about molecules in my transformation into a molecular physicist.

Outside of physics, Durham University volleyball club has provided a much-needed distraction throughout my 8 years at Durham. I must thank all my friends and teammates who made the club such a fun and inclusive environment. Without the many special members of the club I wouldn't have survived a Ph.D. coupled with our almost daily training schedule. Looking back, I am immensely grateful that I was able to play with such a fantastic bunch of athletes and I will always have fond memories of our many achievements.

I would not be where I am today without the support of my whole family. I would especially like to thank my Mam and Dad for always pushing me forward in my endeavours and never letting me forget the important things in life.

Finally, I must thank Steph for her endless love and support.

Contents

Chapter 1
Introduction

1.1 Motivation

Ultracold polar molecules are an exciting avenue of research with potential applications in many areas of physics and chemistry [7, 13, 15, 66]. The richness in applications of ultracold molecules stems from their complex internal structure which allows the study of ultracold chemistry [50], few-body physics [77], quantum simulation of many-body systems [61], quantum computation [24] and the exploration of fundamental physics through precision measurement [15, 79].

In precision measurement with molecules, the most high profile experiments are those involved in the search for the electron electric dipole moment (eEDM), which may help solve the mystery of the baryon asymmetry of the universe by finding on what scale time-reversal symmetry is violated. The search for an eEDM is pursued in multiple experiments [4, 12, 39] using high-mass diatomic molecules (and molecular ions) with a large internal electric field. Cooling of these molecules allows longer coherence times which allow the uncertainties on the value of the eEDM to be pushed to smaller values, placing more stringent constraints on beyond standard model theories. In addition to eEDM experiments, molecules offer other avenues for the investigation of fundamental physics. For example, the vibrational and rotational structure of molecules is sensitive to the variation of fundamental constants, such as the fine structure constant and proton-electron mass ratio; the time-variation of these fundamental constants are of great interest in astrophysics [15, 94]. In addition, molecular spectroscopy may search for physics beyond the standard model by probing the parity violating effects [11] that arise from the anapole moment of the nucleus.

Ultracold molecules provide a testbed for quantum chemistry [50, 77] whereby preparation of internally and externally cold molecules allows chemical reactions to be controlled and probed precisely. Using ultracold molecules it has become possible to observe chemical reactions on the smallest energy scales [69, 71], where quantum effects begin to dominate the outcome of collisions. Already, investigation of quantum chemistry with ultracold molecules that are stable against exchange reactions has raised interesting new questions as to collision processes in these systems; so

© Springer Nature Switzerland AG 2019

A. Guttridge, *Photoassociation of Ultracold CsYb Molecules and Determination of Interspecies Scattering Lengths*, Springer Theses, https://doi.org/10.1007/978-3-030-21201-8_1

called 'sticky collisions' [58] have been proposed as the origin of the unexpected loss process observed in experiments.

Arrays of ultracold molecules provide a new platform for quantum computation and quantum information [24]. The electric dipole moment of heteronuclear molecules gives rise to long-range dipole-dipole interactions. The dipole-dipole interactions between neighbouring molecules may be switched 'on' and 'off' through selective excitation to different internal states [101]. A superposition of molecules in different hyperfine states within the same rotational level are weakly interacting and therefore can act as a storage qubit in a quantum memory [73]. This storage qubit may then be transformed into a strongly-interacting qubit by excitation of the molecule to another rotational level [61].

Polar molecules trapped in optical lattices promise to be a new tool for understanding strongly correlated many-body systems [7, 61]. The long-range dipole-dipole interactions paired with the large number of internal states allow molecules confined in an optical lattice to investigate exotic quantum phases of dipolar quantum gases [33] such as spin ice [31], quantum glasses [53] and supersolid phases [17]. Control of long-range interactions using microwave transitions to different rovibrational states allows the simulation of numerous spin-lattice models [61].

1.2 Cooling Molecules

With the multitude of interesting applications for ultracold molecules there are many experiments worldwide which aim to tackle the significant challenges in trapping and cooling these complex systems. The optimal approach to achieve this goal is far from settled, with a wide range of methods being employed. Because of the wide variation in techniques used, we broadly classify the techniques into two categories: direct and indirect cooling of molecules.

1.2.1 Direct Cooling of Molecules

The direct cooling of molecules follows a predominantly similar approach to the production of a magneto-optical trap (MOT) of atoms. Initially, a source of molecules are produced in a thermal beam, to which slowing techniques are applied to reduce their initial velocity and to allow confinement of the molecules in a trap. The extension of a MOT to molecules is hampered by the same rich internal structure that gives rise to the many exciting properties of molecules. The many internal states provide numerous decay channels from the MOT cooling transition, allowing the molecule to populate a dark state. A major difficulty in creating a molecular MOT is scattering enough photons to sufficiently cool the molecules before they decay to a dark state.

As laser cooling and slowing techniques are not as effective for molecules, it is essential to start off with a cold source of molecules. The development of buffer gas cooling [41, 57, 95] was a critical step in the production of bright, cold molecular beams, providing a large improvement over the much older supersonic expansion technique [45]. Buffer gas cooling involves the use of a buffer gas, such as Helium, flowing through the source chamber. The collisions of the buffer gas with the molecules produces a cold molecular beam which may then be decelerated using a number of techniques.

One of the most common deceleration techniques is Stark deceleration [5, 38, 59, 60, 89], which uses rapidly varying electric fields to slow the molecules. A somewhat less common technique is the magnetic analogue of the Stark decelerator, the Zeeman decelerator [52, 67, 92, 93], which can be used to slow molecules that possess a magnetic moment. The experimental implementation of a Zeeman decelerator is technically more complex than a Stark decelerator due to the difficulty in switching high currents as opposed to high voltages. In recent years there has been significant development of more novel approaches to deceleration of molecules. Such as, deceleration using the centrifugal force [14, 100], laser deceleration using an optical frequency comb [43] and Zeeman-Sisyphus slowing [26], which employs both electromagnetic radiation and magnetic fields similar to an atomic Zeeman slower.

Following the slowing stage, the molecules may then be confined in a trapping potential. Trapping has been demonstrated in an electrostatic trap [6] and an opto-electric trap [100]. Molecular ions have been confined in a trap by reacting a neutral molecule with a trapped atomic ion [16]. Still, the most spectacular recent progress has been in trapping of molecules in a MOT. This feat has now been achieved in SrF [40, 83], CaF [49, 96] and YO [18] systems. The applicability of a molecule to laser cooling relies on highly diagonal Franck-Condon factors (FCF) that allow the scattering of many more photons on the MOT cooling transition before decay to another state. The decay channels to the other states may then be plugged by an array of repump lasers, the number of which depends on the branching ratios specific to the molecule. Once confined in a MOT, these molecules may be cooled into the ultracold regime using blue detuned molasses [25, 96], in a similar scheme to a subset of alkali MOTs. The low temperatures achievable are promising for applications in precision measurement, such as the measurement of the eEDM.

A further challenge in these systems is achieving the low temperatures required to observe molecular collisions or, at even lower temperatures, new quantum phases. The current densities are not sufficient for evaporative cooling in a single sample but sympathetic cooling with an atomic coolant [55], similar to work with molecular ions [80] is a promising approach. A disadvantage specific to the molecular MOT is the MOT's limited applicability to a subset of molecules with highly diagonal FCFs. In this regard, the slowing and trapping techniques mentioned earlier have a much wider applicability, albeit at higher temperatures, for the study of quantum chemistry with 'chemically interesting' molecules. Study of chemically interesting molecules such as CO or OH may aid the understanding of reaction processes in stars and the interstellar medium.

1.2.2 Indirect Cooling of Molecules

The indirect cooling of molecules is an extremely successful approach for the creation of high phase-space density samples of ultracold molecules. The indirect approach produces ultracold molecules by forming them from laser-cooled atoms. The molecules inherit the high phase-space density of the initial atomic mixture, with very little heating from the association process. This allows the great progress sustained in the cooling of atomic systems to ultracold temperatures to be transferred into the molecular regime.

The first ultracold molecules were produced in high-lying vibrational levels near threshold using photoassociation [44]. In this optical association technique a colliding atom pair is excited to a bound molecular state by absorption of a resonant photon. By careful (or fortunate) choice of photoassociation transition, the excited molecule may decay to a deeply bound level of the electronic ground state. Extraordinarily, this technique has been used to produce molecules in the rovibrational ground state [22, 98, 102]. Notably in the case of the LiCs molecule, the very small Franck-Condon overlap between the scattering state and electronically excited state was enhanced by a broad Feshbach resonance. The existence of this resonance was critical for the production ground state molecules [23]. While this was fortunate for the LiCs system, it illustrates a drawback of the photoassociation technique for the production of ultracold ground-state molecules, which is that the technique relies on the unalterable molecular structure and is only suitable to a small number of systems.

The alternative approach is use to use magnetoassociation across a Feshbach resonance to produce molecules in a high-lying vibrational level of the electronic ground state. Due to the coherent nature of magnetoassociation, the molecules can be controllably associated and dissociated with near unit efficiency when confined in a lattice [19, 91]. These weakly-bound molecules may then be transferred to the rovibrational ground state using stimulated Raman adiabatic passage (STIRAP), with single-pass efficiencies exceeding 90% [64]. The combination of magnetoassociation and STIRAP is an extremely successful method for producing homonuclear [21, 51] and heteronuclear [35, 63, 68, 72, 78, 88] molecules in the lowest rovibrational level of a molecular potential.

1.3 Why CsYb Molecules?

To date, the only ultracold heteronuclear molecules formed in the rovibrational ground state are diatomics composed of two alkali-metal atoms. While these molecules possess an electric dipole moment in their spin-singlet electronic ground state,[1] the orientation of the electron spins in this state means no magnetic dipole moment arises from electronic spin.[2] However, molecules with magnetic and electric

---

[1] When the rovibrational ground state is mixed with a higher rotational state.

[2] There is, however, a very small magnetic moment due to nuclear spin.

dipole moments have been produced in the NaLi system by transferring the molecule to the triplet ground state [78]. The magnetic dipole moment in this system arises because the molecule populates the lowest rovibrational level of the $a^3\Sigma^+$ state, where the electron spins are aligned. Even though the molecules are not in the absolute ground state and can undergo decay, the NaLi molecules exhibit a lifetime of 4.6 s.

In contrast to bi-alkali systems, heteronuclear molecules formed from an alkali atom and a closed-shell atom have a single electronic ground state potential $^2\Sigma$. As the closed-shell atoms are divalent, the pairing of a monovalent alkali atom and closed-shell atom produces a molecule that possesses a magnetic dipole moment due to the remaining unpaired electron spin. Molecules composed of an alkali and a closed-shell atom possess both an electric dipole moment and a magnetic dipole moment in the absolute ground state.

The presence of a magnetic dipole moment in CsYb, in addition to an electric dipole moment, provides an extra degree of tunability to the system allowing a variety of more complex spin Hamiltonians to be simulated [61] and new quantum phases to be explored [75]. In addition, there are further prospects for quantum computation specific to paramagnetic molecules [37, 46]. Controllable quantum chemistry may be achieved in these molecules, whereby collisions between $^2\Sigma$ molecules are manipulated by applied magnetic and electric fields, with the potential to shield molecular collisions [1, 76]. The sensitivity of paramagnetic molecules to both electric and magnetic fields opens up applications in sensitive imaging of electromagnetic fields [2].

Amongst the many alkali-closed-shell mixtures pursued [27, 36, 42, 74, 90, 97, 99], the majority of systems use alkali atoms paired with ytterbium (Yb). The main attraction of using Yb is that it possesses a large number of stable isotopes with different spin statistics. Five of the isotopes are composite bosons and two isotopes are composite fermions. Six of the seven isotopes have been cooled to quantum degeneracy [29, 30, 84–87] meaning that a Mott insulator state is achievable for a range of bosonic and fermionic isotopes. The achievement of unitary lattice filling before association has been demonstrated to be a highly efficient method of molecule formation [19, 20, 51, 65]. In any case, a lattice will be required not only for the investigation of correlated many-body systems but also to protect CsYb from energetically allowed exchange reactions $CsYb + CsYb \rightarrow Cs_2 + Yb_2$, which would otherwise lead to fast loss.

The large mass of Cs coupled with these seven stable Yb isotopes provides large reduced mass tuning of the background scattering length, which increases the likelihood of finding a mixture with favourable properties for molecule formation. Among the alkali-Yb combinations considered, CsYb is predicted to be the most favourable system for the observation of novel Feshbach resonances [9], which have been recently observed in the similar RbSr system [3].

Aside from production of ultracold paramagnetic molecules, the unique atomic mixture of Cs and Yb provides an appealing platform for the investigation of novel quantum phases [32, 54, 56, 62, 70, 82, 103]. Implementation of a Cs-blind lattice will allow the realisation of an Yb Bose gas pinned in a lattice surrounded by superfluid Cs. This system will allow the study of impurities in a Bose gas [48], polarons [8] and unexplored lattice cooling schemes [34]. Alternatively, using an Yb-blind lattice with a fermionic Yb isotope will allow the investigation of heavy impurities in a Fermi sea [81].

For all the applications described above, knowledge of the Cs+Yb interspecies scattering length will be essential. Both the realisation of a quantum degenerate mixture of Cs+Yb and the prediction of interspecies Feshbach resonances that allow magnetoassociation of CsYb molecules, rely on accurate knowledge of the interspecies interactions. Unfortunately, current ab initio methods for calculating the CsYb interaction potential are not sufficient to predict the interspecies scattering lengths to the desired precision. Therefore, the scattering lengths of Cs+Yb must be measured experimentally.

1.4 Thesis Outline

At the beginning of the work presented in this thesis, the vacuum system and laser systems for both species had been developed and reliable loading of Cs and ^{174}Yb MOTs had been achieved [10, 28, 47]. In this work, we focus on the development of the equipment and techniques required for cooling Cs and Yb to degeneracy and the measurement of the Cs+Yb scattering lengths.

The thesis is structured into the following chapters:

- Chapter 2 explores the various pathways to the rovibrational ground state of CsYb. A brief overview of collisional physics is presented along with a brief description of Feshbach resonances. We describe the various techniques that may be used to produce ultracold molecules and discuss their relevance for CsYb.
- Chapter 3 outlines the experimental apparatus and laser systems used to produce MOTs of Cs and Yb atoms. The 'beam machine' design for Yb fluorescence spectroscopy is described and techniques for stabilising the Yb lasers to bosonic and fermionic isotopes are discussed.
- Chapter 4 describes the production of quantum degenerate gases of Yb. The incorporation of an optical dipole trap into the setup is discussed and techniques for precooling and loading Yb atoms are presented. The first realisations of ^{174}Yb BEC and ^{173}Yb DFG in the experiment are reported.
- Chapter 5 describes the production of a Cs BEC. The implementation of DRSC for the production of high PSD samples of Cs is characterised and the setup of an additional large volume 'reservoir' dipole trap is described. The combination of DRSC and the reservoir and dimple traps are used to produce the first Cs BEC in this apparatus.
- Chapter 6 reports on initial measurements of the CsYb interspecies scattering length using an optically trapped sample of the two species. We derive a rate equation model to describe the observed sympathetic cooling of Cs by Yb and use the model to extract thermalisation cross sections. The cross sections are com-

pared to quantum scattering calculations to allow the prediction of the interspecies scattering lengths.

- Chapter 7 demonstrates the production of Cs*Yb molecules in a dipole trap using one-photon photoassociation. We describe the development of the photoassociation laser system and the initial testing on the Cs_2 system. We report binding energy measurements of rovibrational states in the 2(1/2) molecular potential for four CsYb isotopologs and Le Roy-Bernstein analysis of the measurements.
- Chapter 8 reports the measurement of the Cs+Yb scattering lengths using two-photon photoassociation. We describe the spectroscopy methods used to create the ground state CsYb molecules and present measurements of the bound-bound transition strengths using Autler-Townes spectroscopy.
- Chapter 9 summarises the experiments performed in the previous chapters. Using the knowledge gained from the experiments presented in this thesis we offer an outlook on the future of the experiment.

A broad range of topics in atomic and molecular physics are covered in this thesis. A thorough treatment of each topic broached is beyond the scope of a single PhD thesis. Therefore, a summary of the relevant background theory is outlined at the appropriate points within each chapter and the interested reader is directed to the appropriate references for seminal papers and detailed review articles.

1.5   Contributions of the Author

Complex dual-species experiments often require a team of people both to operate and to advance the experiment. The CsYb experimental apparatus was constructed by PhD students Kirsteen Butler, Stefan Kemp and Ruben Freytag and postdoctoral researcher Stephen Hopkins. Most of the experimental setup described in Chap. 3, the vacuum chamber, magnetic field coils and the majority of the laser systems, were completed before I began work on the project. Of the work presented in Chap. 3, I performed the setup of the 399 nm laser, the setup of the DRSC optics and the Yb fluorescence spectroscopy measurements.

The setup of the dipole trap in Chap. 4 and the investigation of the Yb MOT was conducted by myself, Stefan and Steve. Following our investigation of loading the optical dipole trap Stefan left the lab to write his thesis and the production and measurements of the ^{174}Yb BEC were performed by Steve and I. The production of the ^{173}Yb DFG and the investigation of other Yb isotopes were both performed by myself.

The Cs DRSC and BEC results presented in Chap. 5 were obtained by Steve and I. The initial thermalisation measurements were performed alongside Steve but following his retirement, all the remaining experimental work presented in this thesis was performed by myself. The only exception being the CsYb one-photon photoassociation data which was taken with the aid of John McFerran, a visiting academic from the University of Western Australia.

The CsYb experiment has benefited from a long-standing collaboration with the theory group of Jeremy Hutson. Jeremy and Matthew Frye performed the quantum scattering calculations presented in the Chap. 6. Also, in Chap. 8, Jeremy, Matthew and Baochun Yang performed the calculations of the interaction potential and the subsequent fitting of the potential to the ground state binding energies. The predictions of interspecies Feshbach resonances presented in the outlook were also performed by Jeremy, Matthew and Baochun.

References

1. Abrahamsson E, Tscherbul TV, Krems RV (2007) Inelastic collisions of cold polar molecules in nonparallel electric and magnetic fields. J Chem Phys 127(4):044,302. https://doi.org/10.1063/1.2748770
2. Alyabyshev SV, Lemeshko M, Krems RV (2012) Sensitive imaging of electromagnetic fields with paramagnetic polar molecules. Phys Rev A 86(1):013,409. https://doi.org/10.1103/PhysRevA.86.013409
3. Barbé V, Ciamei A, Pasquiou B, Reichsöllner L, Schreck F, Zuchowski PS, Hutson JM (2018) Observation of Feshbach resonances between alkali and closed-shell atoms. Nat Phys 14:881–884. https://doi.org/10.1038/s41567-018-0169-x
4. Baron J, Campbell WC, DeMille D, Doyle JM, Gabrielse G, Gurevich YV, Hess PW, Hutzler NR, Kirilov E, Kozyrev I, O'Leary BR, Panda CD, Parsons MF, Petrik ES, Spaun B, Vutha AC, West AD (2014) Order of magnitude smaller limit on the electric dipole moment of the electron. Science 343(6168):269–272. https://doi.org/10.1126/science.1248213
5. Bethlem HL, Berden G, Meijer G (1999) Decelerating neutral dipolar molecules. Phys Rev Lett 83(8):1558–1561. https://doi.org/10.1103/physrevlett.83.1558
6. Bethlem HL, Berden G, Crompvoets FMH, Jongma RT, van Roij AJA, Meijer G (2000) Electrostatic trapping of ammonia molecules. Nature 406(6795):491–494. https://doi.org/10.1038/35020030
7. Bohn JL, Rey AM, Ye J (2017) Cold molecules: progress in quantum engineering of chemistry and quantum matter. Science 357(6355):1002–1010. https://doi.org/10.1126/science.aam6299
8. Bruderer M, Klein A, Clark SR, Jaksch D (2008) Transport of strong-coupling polarons in optical lattices. New J Phys 10(3):033,015. https://doi.org/10.1088/1367-2630/10/3/033015
9. Brue DA, Hutson JM (2013) Prospects of forming ultracold molecules in $^2\Sigma$ states by magnetoassociation of alkali-metal atoms with Yb. Phys Rev A 87(5):052,709. https://doi.org/10.1103/physreva.87.052709
10. Butler KL (2014) A dual species MOT of Yb and Cs. PhD thesis
11. Cahn SB, Ammon J, Kirilov E, Gurevich YV, Murphree D, Paolino R, Rahmlow DA, Kozlov MG, DeMille D (2014) Zeeman-tuned rotational level-crossing spectroscopy in a diatomic free radical. Phys Rev Lett 112(16):163,002. https://doi.org/10.1103/PhysRevLett.112.163002
12. Cairncross WB, Gresh DN, Grau M, Cossel KC, Roussy TS, Ni Y, Zhou Y, Ye J, Cornell EA (2017) Precision measurement of the electron's electric dipole moment using trapped molecular ions. Phys Rev Lett 119(15):153,001. https://doi.org/10.1103/physrevlett.119.153001
13. Carr LD, DeMille D, Krems RV, Ye J (2009) Cold and ultracold molecules: science, technology and applications. New J Phys 11(5):055,049. https://doi.org/10.1088/1367-2630/11/5/055049
14. Chervenkov S, Wu X, Bayerl J, Rohlfes A, Gantner T, Zeppenfeld M, Rempe G (2014) Continuous centrifuge decelerator for polar molecules. Phys Rev Lett 112(1):013,001. https://doi.org/10.1103/PhysRevLett.112.013001

15. Chin C, Flambaum VV, Kozlov MG (2009) Ultracold molecules: new probes on the variation of fundamental constants. New J Phys 11(5):055,048. https://doi.org/10.1088/1367-2630/11/5/055048

16. Chou CW, Kurz C, Hume DB, Plessow PN, Leibrandt DR, Leibfried D (2017) Preparation and coherent manipulation of pure quantum states of a single molecular ion. Nature 545(7653):203–207. https://doi.org/10.1038/nature22338

17. Cinti F, Macrì T, Lechner W, Pupillo G, Pohl T (2014) Defect-induced supersolidity with soft-core bosons. Nat Commun 5:3235. https://doi.org/10.1038/ncomms4235

18. Collopy AL, Ding S, Wu Y, Finneran IA, Anderegg L, Augenbraun BL, Doyle JM, Ye J (2018) 3-D magneto-optical trap of yttrium monoxide. Phys Rev Lett 121(21):213201 https://doi.org/10.1103/PhysRevLett.121.213201

19. Covey JP, Moses SA, Gärttner M, Safavi-Naini A, Miecnikowski MT, Fu Z, Schachenmayer J, Julienne PS, Rey AM, Jin DS, Ye J (2016) Doublon dynamics and polar molecule production in an optical lattice. Nat Commun 7(11):279. https://doi.org/10.1038/ncomms11279

20. Danzl JG, Haller E, Gustavsson M, Mark MJ, Hart R, Bouloufa N, Dulieu O, Ritsch H, Nägerl HC (2008) Quantum gas of deeply bound ground state molecules. Science 321(5892):1062–1066. https://doi.org/10.1126/science.1159909

21. Danzl JG, Mark MJ, Haller E, Gustavsson M, Hart R, Aldegunde J, Hutson JM, Nägerl HC (2010) An ultracold high-density sample of rovibronic ground-state molecules in an optical lattice. Nat Phys 6(4):265–270. https://doi.org/10.1038/nphys153

22. Deiglmayr J, Grochola A, Repp M, Mörtlbauer K, Glück C, Lange J, Dulieu O, Wester R, Weidemüller M (2008) Formation of ultracold polar molecules in the rovibrational ground state. Phys Rev Lett 101(13):133,004. https://doi.org/10.1103/PhysRevLett.101.133004

23. Deiglmayr J, Pellegrini P, Grochola A, Repp M, Côté R, Dulieu O, Wester R, Weidemüller M (2009) Influence of a Feshbach resonance on the photoassociation of LiCs. New J Phys 11(5):055,034. https://doi.org/10.1088/1367-2630/11/5/055034

24. DeMille D (2002) Quantum computation with trapped polar molecules. Phys Rev Lett 88(6):067,901. https://doi.org/10.1103/PhysRevLett.88.067901

25. Devlin JA, Tarbutt MR (2016) Three-dimensional Doppler, polarization-gradient, and magneto-optical forces for atoms and molecules with dark states. New J Phys 18(12):123,017. https://doi.org/10.1088/1367-2630/18/12/123017

26. Fitch NJ, Tarbutt MR (2016) Principles and design of a Zeeman-Sisyphus decelerator for molecular beams. ChemPhysChem 17(22):3609–3623. https://doi.org/10.1002/cphc.201600656

27. Flores AS, Mishra HP, Vassen W, Knoop S (2017) An ultracold, optically trapped mixture of ^{87}Rb and metastable ^{4}He atoms. Eur Phys J D 71(3):49. https://doi.org/10.1140/epjd/e2017-70675-y

28. Freytag R (2015) Simultaneous magneto-optical trapping of ytterbium and caesium. PhD thesis

29. Fukuhara T, Sugawa S, Takahashi Y (2007a) Bose-Einstein condensation of an ytterbium isotope. Phys Rev A 76(5):051,604. https://doi.org/10.1103/PhysRevA.76.051604

30. Fukuhara T, Takasu Y, Kumakura M, Takahashi Y (2007b) Degenerate Fermi gases of ytterbium. Phys Rev Lett 98(3):030,401. https://doi.org/10.1103/PhysRevLett.98.030401

31. Glaetzle AW, Dalmonte M, Nath R, Rousochatzakis I, Moessner R, Zoller P (2014) Quantum spin-ice and dimer models with Rydberg atoms. Phys Rev X 4(041):037. https://doi.org/10.1103/PhysRevX.4.041037

32. Günter K, Stöferle T, Moritz H, Köhl M, Esslinger T (2006) Bose-Fermi mixtures in a three-dimensional optical lattice. Phys Rev Lett 96(18):180,402. https://doi.org/10.1103/physrevlett.96.180402

33. Góral K, Santos L, Lewenstein M (2002) Quantum phases of dipolar bosons in optical lattices. Phys Rev Lett 88(170):406. https://doi.org/10.1103/PhysRevLett.88.170406

34. Griessner A, Daley AJ, Clark SR, Jaksch D, Zoller P (2006) Dark-state cooling of atoms by superfluid immersion. Phys Rev Lett 97(22):220,403. https://doi.org/10.1103/PhysRevLett.97.220403

35. Guo M, Zhu B, Lu B, Ye X, Wang F, Vexiau R, Bouloufa-Maafa N, Quéméner G, Dulieu O, Wang D (2016) Creation of an ultracold gas of ground-state dipolar $^{23}Na^{87}Rb$ molecules. Phys Rev Lett 116(205):303. https://doi.org/10.1103/PhysRevLett.116.205303

36. Hara H, Takasu Y, Yamaoka Y, Doyle JM, Takahashi Y (2011) Quantum degenerate mixtures of alkali and alkaline-earth-like atoms. Phys Rev Lett 106(20):205,304. https://doi.org/10.1103/PhysRevLett.106.205304

37. Herrera F, Cao Y, Kais S, Whaley KB (2014) Infrared-dressed entanglement of cold open-shell polar molecules for universal matchgate quantum computing. New J Phys 16(7):075,001. https://doi.org/10.1088/1367-2630/16/7/075001

38. Hoekstra S, Metsälä M, Zieger PC, Scharfenberg L, Gilijamse JJ, Meijer G, van de Meerakker SYT (2007) Electrostatic trapping of metastable NH molecules. Phys Rev A 76(6):063,408. https://doi.org/10.1103/PhysRevA.76.063408

39. Hudson J, Kara D, Smallman I, Sauer B, Tarbutt M, Hinds E (2011) Improved measurement of the shape of the electron. Nature 473(7348):493–496. https://doi.org/10.1038/nature10104

40. Hummon MT, Yeo M, Stuhl BK, Collopy AL, Xia Y, Ye J (2013) 2D magneto-optical trapping of diatomic molecules. Phys Rev Lett 110(14):143,001. https://doi.org/10.1103/PhysRevLett.110.143001

41. Hutzler NR, Lu HI, Doyle JM (2012) The buffer gas beam: an intense, cold, and slow source for atoms and molecules. Chem Rev 112(9):4803–4827. https://doi.org/10.1021/cr200362u

42. Ivanov VV, Khramov A, Hansen AH, Dowd WH, Münchow F, Jamison AO, Gupta S (2011) Sympathetic cooling in an optically trapped mixture of alkali and spin-singlet atoms. Phys Rev Lett 106(153):201. https://doi.org/10.1103/PhysRevLett.106.153201

43. Jayich A, Long X, Campbell W (2016) Direct frequency comb laser cooling and trapping. Phys Rev X 6(4):041,004. https://doi.org/10.1103/physrevx.6.041004

44. Jones KM, Tiesinga E, Lett PD, Julienne PS (2006) Ultracold photoassociation spectroscopy: long-range molecules and atomic scattering. Rev Mod Phys 78(2):483–535. https://doi.org/10.1103/revmodphys.78.483

45. Kantrowitz A, Grey J (1951) A high intensity source for the molecular beam. part i. theoretical. Rev Sci Instrum 22(5):328–332. https://doi.org/10.1063/1.1745921

46. Karra M, Sharma K, Friedrich B, Kais S, Herschbach D (2016) Prospects for quantum computing with an array of ultracold polar paramagnetic molecules. J Chem Phys 144(9):094,301. https://doi.org/10.1063/1.4942928

47. Kemp SL (2017) Laser cooling and optical trapping of ytterbium. PhD thesis, Durham University

48. Klein A, Bruderer M, Clark SR, Jaksch D (2007) Dynamics, dephasing and clustering of impurity atoms in Bose-Einstein condensates. New J Phys 9(11):411. https://doi.org/10.1088/1367-2630/9/11/411

49. Kozyryev I, Baum L, Matsuda K, Augenbraun BL, Anderegg L, Sedlack AP, Doyle JM (2017) Sisyphus laser cooling of a polyatomic molecule. Phys Rev Lett 118(173):201. https://doi.org/10.1103/PhysRevLett.118.173201

50. Krems RV (2008) Cold controlled chemistry. Phys Chem Chem Phys 10(28):4079–4092. https://doi.org/10.1039/B802322K

51. Lang F, Winkler K, Strauss C, Grimm R, Denschlag JH (2008) Ultracold triplet molecules in the rovibrational ground state. Phys Rev Lett 101(13):133,005. https://doi.org/10.1103/PhysRevLett.101.133005

52. Lavert-Ofir E, Gersten S, Henson AB, Shani I, David L, Narevicius J, Narevicius E (2011) A moving magnetic trap decelerator: a new source of cold atoms and molecules. New J Phys 13(10):103,030. https://doi.org/10.1088/1367-2630/13/10/103030

53. Lechner W, Zoller P (2013) From classical to quantum glasses with ultracold polar molecules. Phys Rev Lett 111(185):306. https://doi.org/10.1103/PhysRevLett.111.185306

54. Lewenstein M, Santos L, Baranov M, Fehrmann H (2004) Atomic Bose-Fermi mixtures in an optical lattice. Phys Rev Lett 92(5):050,401. https://doi.org/10.1103/physrevlett.92.050401

55. Lim J, Frye MD, Hutson JM, Tarbutt MR (2015) Modeling sympathetic cooling of molecules by ultracold atoms. Phys Rev A 92(053):419. https://doi.org/10.1103/PhysRevA.92.053419

56. Marchetti FM, Mathy CJM, Huse DA, Parish MM (2008) Phase separation and collapse in Bose-Fermi mixtures with a Feshbach resonance. Phys Rev B 78(13):134,517. https://doi.org/10.1103/physrevb.78.134517

57. Maxwell SE, Brahms N, deCarvalho R, Glenn DR, Helton JS, Nguyen SV, Patterson D, Petricka J, DeMille D, Doyle JM (2005) High-flux beam source for cold, slow atoms or molecules. Phys Rev Lett 95(17):173,201. https://doi.org/10.1103/physrevlett.95.173201

58. Mayle M, Quéméner G, Ruzic BP, Bohn JL (2013) Scattering of ultracold molecules in the highly resonant regime. Phys Rev A 87(1):012,709. https://doi.org/10.1103/physreva.87.012709

59. van de Meerakker SYT, Smeets PHM, Vanhaecke N, Jongma RT, Meijer G (2005) Deceleration and electrostatic trapping of OH radicals. Phys Rev Lett 94(2):023,004. https://doi.org/10.1103/PhysRevLett.94.023004

60. van de Meerakker SYT, Bethlem HL, Vanhaecke N, Meijer G (2012) Manipulation and control of molecular beams. Chem Rev 112(9):4828–4878. https://doi.org/10.1021/cr200349r

61. Micheli A, Brennen G, Zoller P (2006) A toolbox for lattice-spin models with polar molecules. Nat Phys 2(5):341–347. https://doi.org/10.1038/nphys287

62. Mølmer K (1998) Bose condensates and Fermi gases at zero temperature. Phys Rev Lett 80(9):1804–1807. https://doi.org/10.1103/physrevlett.80.1804

63. Molony PK, Gregory PD, Ji Z, Lu B, Köppinger MP, Le Sueur CR, Blackley CL, Hutson JM, Cornish SL (2014) Creation of ultracold $^{87}Rb^{133}Cs$ molecules in the rovibrational ground state. Phys Rev Lett 113(25):255,301. https://doi.org/10.1103/PhysRevLett.113.255301

64. Molony PK, Gregory PD, Kumar A, Le Sueur CR, Hutson JM, Cornish SL (2016) Production of ultracold $^{87}Rb^{133}Cs$ in the absolute ground state: complete characterisation of the STIRAP transfer. ChemPhysChem 17(22):3811–3817. https://doi.org/10.1002/cphc.201600501

65. Moses SA, Covey JP, Miecnikowski MT, Yan B, Gadway B, Ye J, Jin DS (2015) Creation of a low-entropy quantum gas of polar molecules in an optical lattice. Science 350(6261):659–662. https://doi.org/10.1126/science.aac6400

66. Moses SA, Covey JP, Miecnikowski MT, Jin DS, Ye J (2016) New frontiers for quantum gases of polar molecules. Nat Phys 13(1):13–20. https://doi.org/10.1038/nphys3985

67. Narevicius E, Libson A, Parthey CG, Chavez I, Narevicius J, Even U, Raizen MG (2008) Stopping supersonic oxygen with a series of pulsed electromagnetic coils: a molecular coilgun. Phys Rev A 77(5):051,401. https://doi.org/10.1103/physreva.77.051401

68. Ni KK, Ospelkaus S, de Miranda MHG, Pe'er A, Neyenhuis B, Zirbel JJ, Kotochigova S, Julienne PS, Jin DS, Ye J (2008) A high phase-space-density gas of polar molecules. Science 322(5899):231–235. https://doi.org/10.1126/science.1163861

69. Ni KK, Ospelkaus S, Wang D, Quéméner G, Neyenhuis B, De Miranda M, Bohn J, Ye J, Jin D (2010) Dipolar collisions of polar molecules in the quantum regime. Nature 464(7293):1324–1328. https://doi.org/10.1038/nature08953

70. Ospelkaus S, Ospelkaus C, Humbert L, Sengstock K, Bongs K (2006) Tuning of heteronuclear interactions in a degenerate Fermi-Bose mixture. Phys Rev Lett 97(12):120,403. https://doi.org/10.1103/physrevlett.97.120403

71. Ospelkaus S, Ni KK, Wang D, De Miranda M, Neyenhuis B, Quéméner G, Julienne P, Bohn J, Jin D, Ye J (2010) Quantum-state controlled chemical reactions of ultracold potassium-rubidium molecules. Science 327(5967):853–857. https://doi.org/10.1126/science.1184121

72. Park JW, Will SA, Zwierlein MW (2015) Ultracold dipolar gas of fermionic $^{23}Na^{40}K$ molecules in their absolute ground state. Phys Rev Lett 114(205):302. https://doi.org/10.1103/PhysRevLett.114.205302

73. Park JW, Yan ZZ, Loh H, Will SA, Zwierlein MW (2017) Second-scale nuclear spin coherence time of ultracold $^{23}Na^{40}K$ molecules. Science 357(6349):372–375. https://doi.org/10.1126/science.aal5066

74. Pasquiou B, Bayerle A, Tzanova SM, Stellmer S, Szczepkowski J, Parigger M, Grimm R, Schreck F (2013) Quantum degenerate mixtures of strontium and rubidium atoms. Phys Rev A 88(2):023,601. https://doi.org/10.1103/PhysRevA.88.023601

75. Pérez-Ríos J, Herrera F, Krems RV (2010) External field control of collective spin excitations in an optical lattice of $^2\Sigma$ molecules. New J Phys 12(10):103,007. https://doi.org/10.1088/1367-2630/12/10/103007

76. Quéméner G, Bohn JL (2016) Shielding $^2\Sigma$ ultracold dipolar molecular collisions with electric fields. Phys Rev A 93(012):704. https://doi.org/10.1103/PhysRevA.93.012704

77. Quemener G, Julienne PS (2012) Ultracold molecules under control! Chem Rev 112(9):4949–5011. https://doi.org/10.1021/cr300092g

78. Rvachov TM, Son H, Sommer AT, Ebadi S, Park JJ, Zwierlein MW, Ketterle W, Jamison AO (2017) Long-lived ultracold molecules with electric and magnetic dipole moments. Phys Rev Lett 119(143):001. https://doi.org/10.1103/PhysRevLett.119.143001

79. Safronova MS, Budker D, DeMille D, Kimball DFJ, Derevianko A, Clark CW (2018) Search for new physics with atoms and molecules. Rev Mod Phys 90(025):008. https://doi.org/10.1103/RevModPhys.90.025008

80. Schmid PC, Greenberg J, Miller MI, Loeffler K, Lewandowski HJ (2017) An ion trap time-of-flight mass spectrometer with high mass resolution for cold trapped ion experiments. Rev Sci Instrum 88(12):123,107. https://doi.org/10.1063/1.4996911

81. Schmidt R, Knap M, Ivanov DA, You JS, Cetina M, Demler E (2018) Universal many-body response of heavy impurities coupled to a Fermi sea: a review of recent progress. Rep Prog Phys 81(2):024,401. https://doi.org/10.1088/1361-6633/aa9593

82. Sengupta K, Dupuis N, Majumdar P (2007) Bose-Fermi mixtures in an optical lattice. Phys Rev A 75(6):063,625. https://doi.org/10.1103/physreva.75.063625

83. Steinecker MH, McCarron DJ, Zhu Y, DeMille D (2016) Improved radio-frequency magneto-optical trap of SrF molecules. ChemPhysChem 17(22):3664–3669. https://doi.org/10.1002/cphc.201600967

84. Sugawa S, Yamazaki R, Taie S, Takahashi Y (2011) Bose-Einstein condensate in gases of rare atomic species. Phys Rev A 84(1):011,610. https://doi.org/10.1103/PhysRevA.84.011610

85. Taie S, Takasu Y, Sugawa S, Yamazaki R, Tsujimoto T, Murakami R, Takahashi Y (2010) Realization of a SU (2)× SU (6) system of fermions in a cold atomic gas. Phys Rev Lett 105(19):190,401. https://doi.org/10.1103/PhysRevLett.105.190401

86. Takasu Y, Takahashi Y (2009) Quantum degenerate gases of ytterbium atoms. J Phys Soc Jpn 78(1):012,001. https://doi.org/10.1143/JPSJ.78.012001

87. Takasu Y, Maki K, Komori K, Takano T, Honda K, Kumakura M, Yabuzaki T, Takahashi Y (2003) Spin-singlet Bose-Einstein condensation of two-electron atoms. Phys Rev Lett 91(4):040,404. https://doi.org/10.1103/PhysRevLett.91.040404

88. Takekoshi T, Reichsöllner L, Schindewolf A, Hutson JM, Le Sueur CR, Dulieu O, Ferlaino F, Grimm R, Nägerl HC (2014) Ultracold dense samples of dipolar RbCs molecules in the rovibrational and hyperfine ground state. Phys Rev Lett 113(205301):205,301. https://doi.org/10.1103/physrevlett.113.205301

89. Tarbutt MR, Bethlem HL, Hudson JJ, Ryabov VL, Ryzhov VA, Sauer BE, Meijer G, Hinds EA (2004) Slowing heavy, ground-state molecules using an alternating gradient decelerator. Phys Rev Lett 92(17):173,002. https://doi.org/10.1103/PhysRevLett.92.173002

90. Tassy S, Nemitz N, Baumer F, Höhl C, Batär A, Görlitz A (2010) Sympathetic cooling in a mixture of diamagnetic and paramagnetic atoms. J Phys B: At, Mol Opt Phys 43(20):205,309. https://doi.org/10.1088/0953-4075/43/20/205309

91. Thalhammer G, Winkler K, Lang F, Schmid S, Grimm R, Denschlag JH (2006) Long-lived Feshbach molecules in a three-dimensional optical lattice. Phys Rev Lett 96(050):402. https://doi.org/10.1103/PhysRevLett.96.050402

92. Toscano J, Tauschinsky A, Dulitz K, Rennick CJ, Heazlewood BR, Softley TP (2017) Zeeman deceleration beyond periodic phase space stability. New J Phys 19(8):083,016. https://doi.org/10.1088/1367-2630/aa7ef5

93. Trimeche A, Bera MN, Cromières JP, Robert J, Vanhaecke N (2011) Trapping of a supersonic beam in a traveling magnetic wave. Eur Phys J D 65(1–2):263–271. https://doi.org/10.1140/epjd/e2011-20096-1

94. Truppe S, Hendricks R, Tokunaga S, Lewandowski H, Kozlov M, Henkel C, Hinds E, Tarbutt M (2013) A search for varying fundamental constants using hertz-level frequency measurements of cold CH molecules. Nat Commun 4:2600. https://doi.org/10.1038/ncomms3600

95. Truppe S, Hambach M, Skoff SM, Bulleid NE, Bumby JS, Hendricks RJ, Hinds EA, Sauer BE, Tarbutt MR (2017a) A buffer gas beam source for short, intense and slow molecular pulses. J Mod Opt 65(5–6):648–656. https://doi.org/10.1080/09500340.2017.1384516

96. Truppe S, Williams HJ, Hambach M, Caldwell L, Fitch NJ, Hinds EA, Sauer BE, Tarbutt MR (2017b) Molecules cooled below the Doppler limit. Nat Phys 13(12):1173–1176. https://doi.org/10.1038/nphys4241

97. Vaidya VD, Tiamsuphat J, Rolston SL, Porto JV (2015) Degenerate Bose-Fermi mixtures of rubidium and ytterbium. Phys Rev A 92(043):604. https://doi.org/10.1103/PhysRevA.92.043604

98. Viteau M, Chotia A, Allegrini M, Bouloufa N, Dulieu O, Comparat D, Pillet P (2008) Optical pumping and vibrational cooling of molecules. Science 321(5886):232–234. https://doi.org/10.1126/science.1159496

99. Witkowski M, Nagórny B, Munoz-Rodriguez R, Ciuryło R, Zuchowski PS, Bilicki S, Piotrowski M, Morzyński P, Zawada M (2017) Dual Hg-Rb magneto-optical trap. Opt Express 25(4):3165–3179. https://doi.org/10.1364/OE.25.003165

100. Wu X, Gantner T, Koller M, Zeppenfeld M, Chervenkov S, Rempe G (2017) A cryofuge for cold-collision experiments with slow polar molecules. Science 358(6363):645–648. https://doi.org/10.1126/science.aan3029

101. Yelin SF, Kirby K, Côté R (2006) Schemes for robust quantum computation with polar molecules. Phys Rev A 74(050):301. https://doi.org/10.1103/PhysRevA.74.050301

102. Zabawa P, Wakim A, Haruza M, Bigelow NP (2011) Formation of ultracold $X^1\Sigma^+(v'=0)$ NaCs molecules via coupled photoassociation channels. Phys Rev A 84(061):401. https://doi.org/10.1103/PhysRevA.84.061401

103. Zaccanti M, D'Errico C, Ferlaino F, Roati G, Inguscio M, Modugno G (2006) Control of the interaction in a Fermi-Bose mixture. Phys Rev A 74(4):041,605. https://doi.org/10.1103/physreva.74.041605

Chapter 2
Routes to Ground State CsYb Molecules

In recent years, the creation of ultracold heteronuclear molecules in the rovibrational ground state has witnessed spectacular progress, with ground state molecules of KRb [52], RbCs [50, 66], NaK [55] and NaRb [26] produced to date. The recipe for creation of all these bi-alkali molecules is the same: magnetoassociation of an ultracold mixture using an interspecies Feshbach resonance [35], followed by optical transfer to the ground state using stimulated Raman adiabatic passage (STIRAP) [6, 70].

Mixtures of an alkali atom and a closed-shell atom, like CsYb, do not possess broad Feshbach resonances like bi-alkali systems. Therefore, the best route to the ground state remains unknown in these systems. The optimal pathway for CsYb may involve STIRAP transfer from a Feshbach state as in many bi-alkali systems or possibly from two free atoms as in Sr_2 [14, 61]. Alternatively, the best route could involve initial production of electronic ground state molecules through decay from a molecular state which is excited by one-photon photoassociation. Following decay from the electronically excited molecular state, the high-lying electronic ground state molecules may then be transferred to the rovibrational ground state using STIRAP.

In this chapter we discuss the essential physics behind these molecule formation techniques and assess their applicability to forming rovibrational ground state CsYb molecules. Knowledge of the interspecies scattering length is essential for many of these techniques (and for cooling of the atomic mixture), we therefore begin by discussing the scattering length.

2.1 Scattering Length

Knowledge of the atomic collisional properties is essential when working with ultracold gases. For example, elastic collisions are essential for evaporative cooling of trapped atomic gases and the elastic collision rate dictates the final density a species can be cooled to. Knowledge of the elastic and inelastic loss rates are essential to determine if a degenerate gas may be produced.

© Springer Nature Switzerland AG 2019
A. Guttridge, *Photoassociation of Ultracold CsYb Molecules and Determination of Interspecies Scattering Lengths*, Springer Theses, https://doi.org/10.1007/978-3-030-21201-8_2

The scattering properties of an atomic pair may be characterised by a parameter known as the scattering length. The scattering length a_s parametrises the ultracold scattering properties of a system and is connected to the phase shift δ_0 atoms experience in a collision. The s-wave scattering length is defined as

$$a_s = -\lim_{k \to 0} \frac{\tan \delta_0(k)}{k} \tag{2.1}$$

where k is the collision wavevector of the relative motion of the atoms. The value of the scattering length is determined by the atomic interaction potential $V(R)$ at short range. If the long range behaviour of $V(R)$ is dominated by the van der Waals $-C_6 R^{-6}$ term, the scattering length may be well approximated by

$$a = \bar{a} \left[1 - \tan \left(\Phi - \frac{\pi}{8} \right) \right], \tag{2.2}$$

where $\bar{a} = 0.477988 \ldots (2\mu_r C_6 / \hbar^2)^{1/4}$ is the mean scattering length [25], μ_r is the reduced mass and Φ is the Wentzel-Kramers-Brillouin (WKB) phase integral, which may be written in terms of the interaction potential $V(R)$

$$\Phi = \int_{R_{in}}^{\infty} \sqrt{\frac{2\mu_r}{\hbar^2} V(R)} \, dR. \tag{2.3}$$

In this equation the integration is performed between the classical inner turning point of the potential R_{in} and infinity. Equations 2.2 and 2.3 illustrate the tunability of the scattering length with the reduced mass of the system. This reduced mass tuning of the scattering length is a key asset of alkali-Yb systems, as the large number of stable Yb isotopes produces a large range of reduced masses and therefore a range of scattering lengths for different isotopologs. This increases the chance of finding a favourable combination of Cs and Yb for molecule production.

To verify which Cs+Yb combination is optimal, we must first ascertain the scattering lengths through binding energy measurements of the CsYb molecule's ground state vibrational levels. The scattering length of a system is intrinsically linked to the binding energy of the last bound state of the interatomic potential. A negative scattering length corresponds to a virtual bound state above the atomic threshold, this is equivalent to an attractive interaction. When the last bound state is near threshold (small binding energy) the magnitude of the scattering length is very large and the scattering length diverges when the bound state is at threshold (zero binding energy). If the value of the scattering length is zero, there is effectively no atomic interaction (Ramsauer-Townsend effect) due to the destructive interference of the incoming and outgoing wavefunctions. A positive scattering length signifies a bound state below threshold and corresponds to repulsive interactions. The sign of the scattering length is important to determine if a Bose-Einstein condensate of the atomic species is stable [18].

At ultracold temperatures below the p-wave centrifugal barrier, the elastic collision rate of a particular atom pair is governed by the s-wave elastic scattering cross section σ. For identical particles

$$\sigma_{AA} = 8\pi a_{AA}^2. \tag{2.4}$$

However, in two species experiments, *inter*-species collisions must be considered in addition to the *intra*-species collisions that occur amongst the constituent particles of the mixture. For interspecies collisions the particles are non-identical therefore the elastic scattering cross section becomes

$$\sigma_{AB} = 4\pi a_{AB}^2. \tag{2.5}$$

We will see in Chap. 6 how measurements of the interspecies elastic scattering cross section (and therefore scattering length) are performed by observing the sympathetic cooling of Cs by Yb.

2.2 Feshbach Resonances

Feshbach resonances are an essential tool in atomic and molecular physics that allows precise control of the interaction between atoms [12]. They also offer the ability to coherently transfer a colliding atom pair into a bound molecular state through a technique known as magnetoassociation [35] which is essential in the formation of ultracold molecules from ultracold atoms. The exquisite control of the atomic interactions afforded by Feshbach resonances may also be used to create bright solitons [33, 62] or quantum droplets [10, 30], atomic gases which can propagate without dispersing, bound only by the atomic self interaction. In addition, Feshbach resonances were instrumental in the study of the BEC-BCS crossover in Fermi gases [57, 83].

Control over the collisional properties of an atomic system was essential in the cooling to degeneracy of atomic species with unfavourable background scattering lengths [17, 33, 58, 62, 74]. A pertinent example is the case of Cs which has an extremely large background scattering length [7]. A large scattering length is unfavourable for evaporative cooling because of the a^4 scaling of the three-body recombination rate [22] which causes large atom loss for high densities. Achievement of BEC in Cs [74] required careful control of the atomic interactions, this was provided by a wide Feshbach resonance that allowed broad tunability of the scattering length at low magnetic fields [11].

The physical origin of Feshbach resonances may be easily understood by considering a two-channel model, such as the one presented in Fig. 2.1a. We consider two channels corresponding to molecular potential curves V_{ent} and V_c which describe the entrance and closed channels respectively. The entrance channel typically corresponds to the ground state potential of two atoms and therefore at large internuclear distance R the potential asymptotically approaches the ground state of two free atoms.

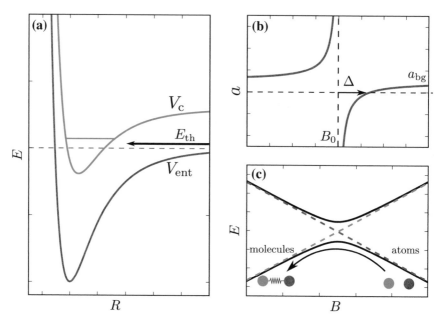

Fig. 2.1 A colliding atom pair with energy E_{th} may couple to a bound state in the closed channel. The bound state in V_c may be tuned relative to the atomic threshold of the entrance channel by use of a magnetic field. **b** As the bound state is tuned towards atomic threshold from higher energy, the scattering length reduces to less than its background value a_{bg} and tends to negative infinity as it approaches the atomic threshold. As the bound state crosses the threshold energy at a magnetic field of B_0, the scattering length becomes positive and reduces in magnitude as the state becomes more bound. **c** The coupling between the molecular and atomic states produces an avoided crossing. An adiabatic sweep of the magnetic field across the avoided crossing transfers the colliding atoms into loosely bound Feshbach molecules

For small collision energies E_{th}, this channel is energetically accessible, hence, the name entrance channel. In contrast, V_c is not energetically accessible and is called the closed channel. However, V_c does support bound molecular states near the threshold of the entrance channel. A Feshbach resonance occurs when a bound molecular state in the closed channel energetically approaches the (free-atom) scattering state in the entrance channel. The energy difference between the two potentials may be tuned using a magnetic field via the Zeeman effect, as the entrance and closed potentials typically have different magnetic moments due to the orientation of the electronic spins. In the case of alkali atoms, the hyperfine interaction gives rise to many additional channels creating many crossings as a function of magnetic field. In the presence of coupling between the two potentials, strong mixing of the channels occurs.

The behaviour of the scattering length close to a magnetically tuned Feshbach resonance may be described by [49]

$$a(B) = a_{bg}\left(1 - \frac{\Delta}{B - B_0}\right),$$

(2.6)

where a_{bg} is the background scattering length, the value of the scattering length away from resonance, B is the magnetic field strength, B_0 is the resonance position and Δ is the resonance width. This expression for the scattering length is illustrated in Fig. 2.1b. The resonance width is defined as the distance between the resonance pole and the zero crossing of the scattering length. The width is determined by the coupling between the two channels and the differential magnetic moment, $\delta \mu = \mu_{atoms} - \mu_{mol}$.

The near-threshold levels that produce the Feshbach resonances are labelled by quantum numbers $n(f_1 f_2)F\ell(M_F)$. F is the resultant of the hyperfine quantum number of the colliding atoms f_1 and f_2 and $M_F = m_1 + m_2$. n is the vibrational level of the molecular potential curve labelled with respect to dissociation, with $n = -1$ denoting the last bound state. The partial wave angular momentum ℓ, may take values of $\ell = 0, 2, 4$ etc. in the bosonic case, these are commonly referred to as $s-$, $d-$, $g-$wave resonances.

In mixtures of alkali atoms the coupling between the channels may occur through strong isotropic electronic interactions which couple states with the same M_F quantum number. The magnitude of the coupling is the difference in energy between the singlet and triplet curves, which is typically large. Coupling may also exist due to spin-spin interactions (magnetic dipole interaction of the spins) which may couple different partial waves ℓ or M_F.

2.2.1 Scattering Properties of Cs

A plot of the Cs $|F = 3, m_F = +3\rangle$ scattering length as a function of magnetic field is shown in Fig. 2.2a. The binding energies of molecular states contributing to the Feshbach resonances in Fig. 2.2a are shown as a function of magnetic field in Fig. 2.2b. Feshbach resonances in the upper plot occur at the magnetic field value where the molecular state crosses threshold ($E_b = 0$). In Cs, multiple Feshbach resonances occur at low magnetic fields due to $\ell = 2$ and $\ell = 4$ molecular states. These higher partial wave Feshbach resonances have observable widths at low magnetic fields due to the strong second order spin-orbit coupling term in Cs [37].

The scattering length of Cs is dominated at low fields by a broad s-wave Feshbach resonance at -12 G [7]. The negative value of the magnetic field corresponds to changing the sign of all the spin projection quantum numbers and is equivalent to stating that there exists a broad Feshbach resonance at 12 G in the $|F = 3, m_F = -3\rangle$ state. The near threshold bound state that produces this Feshbach resonance gives Cs a large scattering length at most magnetic fields.

The broad tunability of the Cs scattering length in the low field region was essential for the creation of a Cs BEC [74]. For high atomic densities the dominant loss process in the $|F = 3, m_F = +3\rangle$ state is three-body loss. Three-body loss occurs when two colliding atoms form a molecule while interacting with a third atom. In the three-body collision the molecular binding energy is released as kinetic energy, typically causing the loss of all three atoms from the trap. The three-body recombination rate scales with scattering length as $\propto a^4$ [75], yielding a very large value for Cs at zero

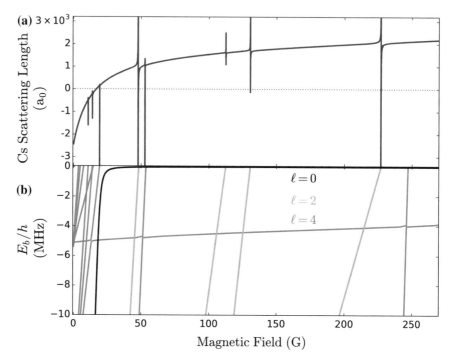

Fig. 2.2 a Cs scattering length as a function of magnetic field for the $|F = 3, m_F = +3\rangle$ state. Several d-wave and g-wave resonances are visible. **b** Molecular binding energies of molecular states as a function of magnetic field. The colours correspond to the rotational angular momentum of the molecular state with black corresponding to $\ell = 0$, green to $\ell = 2$ and blue to $\ell = 4$

magnetic field. Fortunately, there exists an Efimov minimum in the three-body loss rate at $a_{Cs} = 210\,a_0$ [38]. The scattering length can be easily tuned to this value by using a magnetic field around 22 G. At this scattering length, Cs can be evaporatively cooled efficiently due to the large elastic scattering cross section and the suppressed inelastic loss rate.

2.2.2 Magnetoassociation

The association of molecules by tuning the magnetic field near a Feshbach resonance is often called magnetoassociation. Ultracold molecules produced through magnetoassociation are often referred to as Feshbach molecules. The most popular method of magnetoassociation of an atomic pair is by dynamic sweeps of the magnetic field across the resonance pole B_0 [35]. This process is illustrated in Fig. 2.1c. Magnetoassociation by adiabatic ramps of magnetic field was first proposed in a number of publications [1, 48, 68] and has been a very successful technique for the production of heteronuclear molecules from an ultracold atomic mixture [27, 36, 52, 65, 72, 73, 76]. The Feshbach molecules produced retain the high phase space density of the original atomic mixture and occupy a single rovibrational level. Due to the fact that Feshbach resonances occur when molecular and atomic states cross, the Feshbach molecules are formed in a high-lying level near atomic threshold. Therefore, the Feshbach molecules undergo rapid vibrational relaxation due to atom-molecule and molecule-molecule collisions, limiting the lifetime of the molecules to relatively short time scales. For fermionic molecules, the lifetime of the weakly-bound Feshbach molecules are extended due to the suppression of collisions by Pauli blocking [56].

Collisions, along with the phase-space densities and overlap of the two species, typically limit the efficiency of molecule formation in an optical dipole trap, with the best efficiencies approaching 25% [81]. More efficient formation of Feshbach molecules may be achieved in an optical lattice with one atom of each species per lattice site [67]. In a single lattice cell the overlap of the atomic wavefunctions is near unity, which gives the association technique a theoretical efficiency of 100%. Confinement in a lattice also protects the resultant Feshbach molecules from collisions with atoms or other molecules, significantly prolonging their lifetime. The collisional stability allows longer, more adiabatic ramps across the Feshbach resonance increasing the number of molecules produced. Unity conversion efficiencies have been reported for doubly occupied sites in experiments on KRb molecules [13]. In a lattice, the typical efficiency of molecule formation is not limited by that of the association process but by the number of atoms loaded into doubly occupied sites.

Once Feshbach molecules have been formed, the usual next step is to transfer the molecules into the rovibrational ground state using STIRAP. The efficiency of the transfer to the ground state may be improved by using a different initial molecular state. Once a Feshbach molecule is produced, the many avoided crossings of molecular levels illustrated in Fig. 2.2b may be used to transfer the molecule into another state. This can be achieved by controlling the Landau-Zener tunnelling at avoided crossings using magnetic field ramps [46] or using RF transitions [39].

2.3 Feshbach Resonances in Mixtures of Alkali and Closed-Shell Atoms

The difficulty in forming molecules in systems like CsYb comes from the absence of electronic spin in the ground state of the closed-shell atom. In the simple picture considered in the previous section, multiple electronic states occur near atomic threshold due to the interaction of the electronic spins. In the alkali case, this produces a deeply bound singlet and a weakly bound triplet potential. However, when one of the atoms in the colliding pair is spin singlet ($^1S_0$), like Yb, the molecule has only a single electronic state. The molecular levels lie parallel to atomic thresholds as a function of magnetic field, therefore, crossings of atomic and molecular states will only occur between the different hyperfine manifolds of the molecule and alkali atom. In addition, the coupling mechanisms considered in the previous section are not present for CsYb. This is again because of the absence of spin in the Yb ground state.

2.3.1   Ground State Atoms

In the absence of strong coupling between the open and closed channels, a Feshbach resonance will have very small width and not be suitable for magnetoassociation. This is certainly the case for pairs of $^1S_0$ atoms and was also believed to be the case in mixtures of alkali atoms and closed-shell atoms. However, in Ref. [82] the authors introduced a previously unconsidered coupling mechanism which arises due to the modification of the hyperfine coupling constant of the alkali atom by the closed-shell atom at short range. The system considered in the paper was Rb+Sr but the coupling mechanism exists (with varying strength) for all alkali-closed-shell mixtures.

In all cases this mechanism only couples states of equal m_F and ℓ. This selection rule means Feshbach resonances will only occur at crossings between an atom in the lower hyperfine manifold and a molecular state in the upper hyperfine manifold because the magnetic moment of the atomic and molecular states are equal for levels in the same hyperfine manifold. Therefore, only relatively few crossings actually lead to resonances, making the Feshbach spectrum very sparse as a function of magnetic field. In addition, the width of the resonances are narrow ($\lesssim 1$ mG) due to the weak coupling mechanism.

Later studies by Brue and Hutson identified a second coupling mechanism (Mechanism II) that could lead to broader Feshbach resonances in fermionic closed-shell atoms due to the presence of nuclear spin [8]. The coupling occurs between the nuclear spin of the closed-shell atom and the valence electron of the alkali atom. This coupling conserves ℓ and $m_F + m_I$ which leads to the selection rule $\Delta m_F = m_{F,\text{atom}} - m_{F,\text{mol}} = 0, \pm 1$. This relaxed selection rule leads to resonances at many more crossings, including crossings between atomic and molecular states of the same hyperfine manifold.

The prospects for many different alkali-Yb combinations were considered in Ref. [9]. Combinations with Yb were considered because of the large number of isotopes (including two fermions), which gives a considerable reduced mass tuning of the binding energies of the molecular state. Therefore, it increases the likelihood of a Feshbach resonance occurring at a more favourable magnetic field for a specific isotopolog. Of all the alkali-Yb combinations, CsYb was considered as the most promising due to the large reduced mass of the system and the strong hyperfine coupling in Cs. A large reduced mass facilitates a higher density of vibrational levels near threshold, increasing the number of crossings as a function of magnetic field. The large nuclear spin and stronger modification of the hyperfine coupling constant of Cs produces resonances with larger width.

Although the resonances for CsYb are predicted to have larger widths than other alkali-Yb combinations, the predicted widths are still fairly narrow in comparison to bi-alkali Feshbach resonances. Coupled with the knowledge that such resonances will be sparsely distributed as a function of magnetic field, their observation will be extremely challenging without theoretical guidance. The binding energies of the molecular potential must be known to a high precision for accurate predictions of the positions of Feshbach resonances. These must be measured experimentally as even the most sophisticated ab initio calculations are not precise enough to determine the depth of a molecular potential to the required accuracy. An example of the sensitivity of these potentials to the scattering length (binding energy of last bound state) is that a change in the potential depth of $\sim 3\%$ corresponds to a change in the scattering length from $a = -\infty \rightarrow +\infty$. To enable accurate predictions of Feshbach resonance positions, we perform two-photon photoassociation measurements in Chap. 8 to measure the binding energies of the near threshold states of CsYb precisely.

Observation of Feshbach Resonances in RbSr

During the measurements of the two-photon photoassociation features presented later in this thesis (Chap. 8) exciting results were published by Florian Schreck's group in Amsterdam [5]. Working with Rb−Sr mixtures they reported the first observation of Feshbach resonances between alkali and closed-shell atoms. Surprisingly, in mixtures of Rb with the fermionic Sr isotope, ^{87}Sr, more resonances were measured than originally anticipated.

The vast majority of resonances observed were produced through Mechanism II, which shows that mixtures involving fermionic isotopes are favourable for the observation of interspecies resonances. An unexpected result of this investigation was the observation of a series of resonances not initially predicted. The mechanism that produces these resonances (Mechanism III) arises due to the anisotropic interaction of the (shared) electron spin with the nucleus of either the fermionic closed-shell atom (Mechanism IIIa) or the alkali atom (Mechanism IIIb). Therefore, there are two distinct cases: when the electron spin is coupled to the nuclear spin of the closed-shell atom and when the electronic spin couples to the nuclear spin of the alkali atom. Notably, Mechanism IIIb produces additional resonances in bosonic isotopes, although they are predicted to be of much smaller width. Mechanism III conserves $m_F + m_I + m_\ell$ and therefore can couple s-wave states to d-wave ($\ell = 2$) states. The selection rules are different for the two cases, in a: $\Delta m_F = -\Delta m_\ell$ and in b: $\Delta m_F = \pm 1$.

Although Mechanism III produces coupling at many crossings at lower magnetic fields, the strength of the coupling is weaker, resulting in narrower widths. Still, these results seem very promising for the observation of CsYb Feshbach resonances.

2.3.2 Metastable Atoms and Alkali Atoms

In addition to the above resonances, there are other avenues to explore in the search for Feshbach resonances in alkali-Yb systems. Ytterbium possesses many narrow transitions useful for precision spectroscopy, the most celebrated of which is the $^1S_0 \rightarrow {}^3P_0$

transition used in atomic clock experiments all over the globe. The clock transition is extremely narrow as it is doubly forbidden, $\Delta S = 1$ and $J = 0 \rightarrow J' = 0$, meaning excitation to this state yields metastable atoms with a very long lifetime. The lifetime of the state exceeds the background lifetime in many cold atom experiments, making it possible to perform experiments similar to those with ground state atoms by instead using these metastable atoms.

Feshbach resonances have been observed in the fermionic ^{173}Yb($^1S_0$) + ^{173}Yb($^3P_0$) system [28, 54] even though both the ground state and the excited state have zero electronic spin. The resonances arise due to the nuclear spin of the fermionic ^{173}Yb atom and are known as orbital Feshbach resonances because the coupling is provided by the inter-orbital spin-exchange interaction [80]. As electronic spin is absent in both channels, the energy difference of the two channels can only be tuned through the nuclear Zeeman effect which is smaller than the electronic Zeeman effect by $\sim$2000. Fortunately, a molecular bound state lies very close to threshold bringing the resonance within reach of experiments.

Within the same triplet manifold as the $^3P_0$ state is the less explored $^3P_2$ metastable state. The transition from the ground state to this state is $\Delta S = 1$ and $\Delta J = 2$ yielding a long radiative lifetime of around 15 s. As atoms in the $^3P_2$ state are anisotropic, the interaction between the metastable atoms and ground state atoms has several additional coupling mechanisms compared to the interaction between two ground state atoms or between ground state atoms and those in the $^3P_0$ state. Resonances arising from these mechanisms were first observed in a homonuclear mixture of Yb in ground and metastable $^3P_2$ states [31, 63]. Recently Feshbach molecules have been created by association across these Feshbach resonances [64].

The same anisotropic coupling is predicted to occur between alkali atoms and metastable Yb atoms. Work on these novel mixtures has been pursued by groups working on LiYb in Kyoto and Seattle. The metastable Yb atoms may be prepared either by direct excitation on a narrow transition [77] or by excitation to the $^3D_2$ state and the subsequent decay to $^3P_2$ state [34]. Both groups have observed indications of field-dependent inelastic losses [21, 60] but to date no distinct resonances have been found. This is likely due to the suppression of the resonance poles in the presence of inelastic scattering [24, 29], which is prevalent in the Li+Yb system. The presence of large loss rates due to the spin exchange processes make the applicability of these resonances to magnetoassociation questionable. However, these inelastic decays are expected to be weaker for heavier alkali atoms. Currently, this may not be the most promising route for CsYb but this approach may be required if suitable Feshbach resonances with ground state Yb are at magnetic fields too high to reach.

2.4 Photoassociation

In the photoassociation process two colliding atoms A and B are coupled to an electronically excited molecular state by a photon of energy $\hbar \omega_1$. Atoms may be excited to discrete, bound molecular states when the frequency of the laser field, ω_1, is red-detuned from an atomic transition ω_0 and ω_1 is resonant with a vibrational level of the electronically excited molecular potential. On one-photon resonance the colliding atom pair is transferred into a bound electronically excited molecule. The process is illustrated in Fig. 2.3a. The absorption process is given by

$$A + B + \hbar\omega_1 \rightarrow (AB)^*. \tag{2.7}$$

Following excitation, the electronically excited molecular state $(AB)^*$ may then dissociate into two free atoms or decay to the electronic ground state AB. The presence of this decay mechanism offers an alternative route to the creation of ground state molecules. Molecules may be produced in a carefully selected electronically excited state which possesses a favourable wavefunction overlap with the vibrational wavefunction of a rovibrational level in the electronic ground state. This wavefunction overlap, or Franck-Condon factor (FCF), can lead to a large number of molecules accumulating in the ground state due to the decay of the excited molecule. This technique has been used to continuously produce ultracold electronic ground state molecules from an atomic sample [3, 4, 20, 32, 59, 79]. One-photon photoassociation is discussed in much greater depth in Chap. 7.

The electronic ground state molecules produced through decay are preferentially in high-lying vibrational levels. This is required due to the Franck-Condon principle. Typically, producing a large number of molecules in deeply-bound states requires excitation to a excited state vibrational level with a turning point at small interatomic separations ($R < 20\,a_0$). For photoassociation transitions to occur, the interatomic

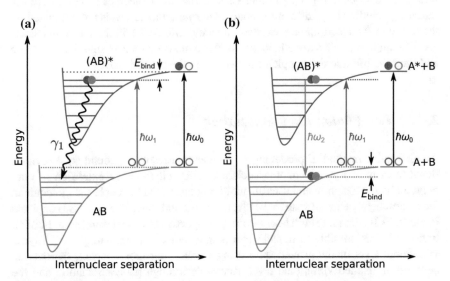

Fig. 2.3 **a** One-photon photoassociation scheme. A colliding atom pair A and B are excited to an electronically excited vibrational level in the $(AB)^*$ potential by a laser of frequency ω_1. **b** Two-photon photoassociation scheme. The electronically excited molecule $(AB)^*$ is coupled to a vibrational level in the electronic ground state by a second laser of frequency ω_2. The internuclear distances at which the transitions occur are not represented to scale

separation of the colliding atom pair must be commensurate to the turning point of the vibrational level. At smaller interatomic distances rapid oscillations of the vibrational wavefunction typically wash out any Franck-Condon overlap. The interatomic separation of the colliding atom pair is dependent on the density but is typically very large ($R > 1000\,a_0$), therefore, the molecular states accessible through photoassociation are those with turning points at large interatomic distances. The long-range molecules in the excited state therefore have very little overlap with more deeply-bound vibrational levels in the electronic ground state which have peaks in the vibrational wavefucntion at shorter range. Therefore, the long-range molecules preferentially decay to the high-lying vibrational levels of the electronic ground state potential. However, deeply bound ground state molecules may be produced through decay in some exceptional systems [20, 53, 79]. Potentials of some alkali dimers, such as Cs_2, exhibit electronically excited molecular potentials with an external long-range well in which a molecule excited to this potential oscillates between long-range and intermediate distances. In this intermediate range the overlap with more deeply bound vibrational levels in the electronic ground state potential is larger. These long-range states in Cs_2 have been studied extensively in an effort to produce ultracold ground state molecules through photoassociation [16, 23, 41–44, 71].

The production of ground state molecules through one-photon photoassociation has some issues. Firstly, the process is incoherent, which makes detection of the molecules more difficult. The molecule cannot be imaged by coherently dissociating it and imaging the free atoms like in magnetoassociation. But, the molecules may be ionised and detected by an micro-channel plate (MCP) [40] or absorption imaged after photodissociation [47]. Another issue is that the molecules produced are not internally 'cold'. Many different ground state rovibrational levels are populated by the decay of the electronically excited molecule. Molecules in the unwanted states are difficult to remove and collisions with these molecules cause significant heating and limit the lifetime the sample of molecules in the desired state.

2.4.1 Two-Photon Photoassociation

An alternative method of producing molecules in a resolved ground state rovibrational level is two-photon photoassociation. This scheme is an extension of one-photon photoassociation and uses an additional laser field to couple the molecule in the electronically excited state to the electronic ground state. The process is depicted in Fig. 2.3b. By detuning both lasers from the respective free-bound and bound-bound transitions it is possible to transfer atoms into molecules populating a high-lying vibrational level of the ground state using Raman transitions. Again, the overlap between an excited vibrational level accessible through photoassociation and the rovibrational ground state is vanishingly small, similar to the one-photon case. This means even with a high intensity laser, appropriately detuned to prevent off-resonant excitation, it is unlikely that the induced coupling will be strong enough for efficient transfer of the population to the ground state. Efficient transfer necessitates a coherent process but Rabi oscillations between atomic and molecular states have (so far) not been observed on an electric-dipole allowed transition due to decoherence arising from spontaneous Raman scattering. The off-resonant scattering of Raman photons is reduced for narrow 'forbidden' transitions where atom-molecule Rabi oscillations have been observed [63, 78]. Transitions to molecular states which dissociate at the Yb ^3P_J atomic threshold are promising for the creation of CsYb molecules using two-photon Raman transitions.

Two-photon transitions using electric-dipole allowed transitions have been used to incoherently produce molecules in the rovibrational ground state [59]. The experiment, performed on RbCs molecules, initially populated a high-lying ground state vibrational level through decay of a photoassociated excited state. The weakly-bound ground state molecules were transferred to the rovibrational ground state using a two-step stimulated emission pumping technique. However, the efficiency of this process is severely limited by the decay of the intermediate excited state.

2.5   STIRAP

The Stimulated emission pumping case discussed previously, whereby two lasers are used to transfer population to a target state via an intermediate state, is one example of how rovibrational ground state molecules may be produced. The issue with this incoherent process is that even when both transitions are saturated (which may be difficult to achieve depending on the Franck-Condon overlap) the transfer is only ~10% efficient [69]. For more efficient transfer between the states, one may utilise a coherent process known as stimulated Raman adiabatic passage (STIRAP). The scheme requires creation of a dark state, which allows STIRAP to be performed using an electric-dipole allowed transition over a time scale longer than the lifetime of the excited state [70].

To explain the STIRAP technique we consider the three-level lambda-type system shown in Fig. 2.4a. We label the initial (Feshbach) state $|F\rangle$, the intermediate excited state $|E\rangle$ and the final (ground) state $|G\rangle$. The laser driving the $|F\rangle \leftrightarrow |E\rangle$ transition is labelled as the 'Pump' laser and for historical reasons the laser driving the $|E\rangle \leftrightarrow |G\rangle$ transition is labelled as the 'Stokes' laser. The equations derived in the following analysis are completely general for any three-level lambda-type system, including the system considered in Chap. 8 in the context of two-photon photoassociation.

The Hamiltonian for this three-level system is [6]

$$\hat{H} = \frac{\hbar}{2} \begin{pmatrix} 0 & \Omega_P(t) & 0 \\ \Omega_P(t) & 2\Delta_P & \Omega_S(t) \\ 0 & \Omega_S(t) & 2(\Delta_P - \Delta_S) \end{pmatrix}, \tag{2.8}$$

where $\Omega_P(t)$, $\Omega_S(t)$ are the Rabi frequencies and $\Delta_{P(S)}$ is the detuning from resonance of the Pump (Stokes) laser. With both lasers on two-photon resonance ($\Delta_P = \Delta_S = 0$), the analytical eigenstates of this Hamiltonian are:

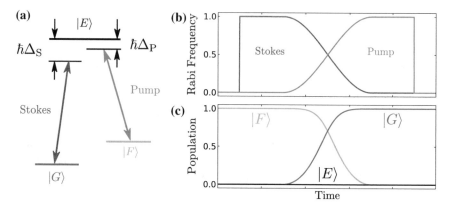

Fig. 2.4 Level scheme and pulse sequence for STIRAP in an ideal three-level system. **a** A three-level lambda-type system. The levels are coupled by two lasers labelled as 'Pump' and 'Stokes'. **b** Pulse sequence for transfer of population from $|F\rangle$ to $|G\rangle$. **c** Temporal evolution of the populations of the three states during the STIRAP pulse sequence

$$|a^+\rangle = \sin\theta\sin\phi|F\rangle + \cos\phi|E\rangle + \cos\theta\sin\phi|G\rangle, \tag{2.9a}$$

$$|a^0\rangle = \cos\theta|F\rangle - \sin\theta|G\rangle, \tag{2.9b}$$

$$|a^-\rangle = \sin\theta\cos\phi|F\rangle - \sin\phi|E\rangle + \cos\theta\cos\phi|G\rangle. \tag{2.9c}$$

Here, we have defined two mixing angles θ and ϕ which are defined by

$$\tan\theta = \frac{\Omega_P}{\Omega_S}, \qquad \tan 2\phi = \frac{\sqrt{\Omega_P^2 + \Omega_S^2}}{\Delta_P}. \tag{2.10}$$

Essential to STIRAP is the dark state $|a^0\rangle$, this state has no component due of intermediate state $|E\rangle$. Using this engineered dark state, 100% transfer from $|F\rangle$ to $|G\rangle$ is possible on timescales longer than the lifetime of $|E\rangle$. The counter-intuitive STIRAP pulse sequence used to achieve efficient transfer is shown in Fig. 2.4b. The sequence is performed by first ramping up the Rabi frequency of the Stokes field Ω_S which prepares for the lossless transfer by Autler-Townes splitting of the levels. Then the Rabi frequency of the pump field Ω_P is ramped up as Ω_S is ramped down which causes the state vector to rotate in the $|F\rangle - |G\rangle$ plane, preventing population of $|E\rangle$. Finally, with the population in $|G\rangle$, Ω_S is reduced to zero and Ω_P can be set to zero following the extinction of the Stokes light.

In the population evolution presented in Fig. 2.4c we see that in this idealised sequence no population is transferred to $|E\rangle$. Not only does the sequence suppress loss from spontaneous emission but it is relatively robust to experimental conditions such as laser intensity and pulse timing. Compare this to a transition driven with an intuitively ordered pulse sequence (Bright STIRAP) which is extremely sensitive to the pulse area due to Rabi flopping of the population. In addition, the timescale of the pulse sequence in Bright STIRAP must be faster than radiative lifetime of $|E\rangle$ to achieve the maximum transfer efficiency of 50% [70]. STIRAP with counter-intuitively ordered pulses is therefore a very effective technique for transfer of the population to the rovibrational ground state.

For transfer of Feshbach molecules to the rovibrational ground state efficiencies of 92% have been achieved [51]. The transfer efficiency is reliant upon maintaining an adiabatic evolution of the dark state. Phase fluctuations between the lasers can cause the state vector to deviate from the $|F\rangle - |G\rangle$ plane, causing non-adiabatic coupling. This process limits the efficiency in realistic systems and may occur on a timescale much longer than the excited state lifetime. The condition for efficient transfer is [2]

$$\frac{\Omega_P^2 + \Omega_S^2}{\pi^2 \gamma} \gg \frac{1}{\tau} \gg D, \tag{2.11}$$

where γ is the linewidth of the intermediate state and D is the linewidth associated with the frequency difference of the lasers.

The job of an experimenter is therefore made easier by choice of an intermediate state with a long lifetime and strong coupling to initial and final states. This places less stringent requirements on the relative linewidth of the STIRAP lasers. However, the choice of transitions is reliant on molecular structure and there is no guarantee that a suitable intermediate state for transfer to the rovibrational ground state exists. In this case it may be more favourable to perform multiple STIRAP pulses, the first from the Feshbach state to a more deeply bound state and then from this state to the ground state. Two possible four-photon schemes: sequential and 'straddle STIRAP', were investigated for the purposes of transferring Cs_2 molecules to the rovibrational ground state [19]. In both schemes a single-pass efficiency of $\simeq 60\%$ was achieved.

2.5.1 Free-Bound STIRAP

Given the high transfer efficiencies possible between bound molecular states using STIRAP, it is reasonable to consider the possibility of using STIRAP to efficiently convert an atom pair into a ground state molecule. Of course, it is highly unlikely that in a single STIRAP sequence the atoms could be transferred to the rovibraitonal ground state but, as demonstrated by Cs_2, four-photon STIRAP may be used to transfer the population to the rovibrational ground state.

The difficulty in Free-Bound STIRAP arises due to the extremely weak free-bound transition. The free-bound transition is typically $10^3 - 10^4$ times weaker than a bound-bound transition due to the unfavourable Franck-Condon overlap. The free-bound Rabi frequency can be enhanced in Bose condensed systems by Bose-enhancement [45]. In this case the free-bound Rabi frequency is enhanced by a factor of $\sqrt{N}$, where N is the number of Bose condensed atoms. This effect was recently observed in the creation of Sr_2 molecules [15].

A more promising approach for Free-Bound STIRAP is performing the pulse sequence on a sample of atoms trapped in a lattice [14, 61]. The advantages of using a lattice are two-fold. Firstly, confinement in a lattice increases the free-bound Rabi frequency by increasing the amplitude of the atomic wavefunction at short range [14]. In the Sr_2 work, the authors found that using a Mott insulator state can increase the free-bound Rabi frequency by a factor of four in comparison to Bose enhancement. Secondly, the lifetime of the molecules is increased by suppression of atom-molecule collisions due to the confinement of one molecule per lattice site. Using a narrow electric-dipole forbidden transition in Sr, 30% of atoms in doubly-occupied sites have been transferred into molecules [14].[1]

References

1. van Abeelen FA, Verhaar BJ (1999) Time-dependent Feshbach resonance scattering and anomalous decay of a Na Bose-Einstein condensate. Phys Rev Lett 83:1550–1553. https://doi.org/10.1103/PhysRevLett.83.1550
2. Aikawa K, Akamatsu D, Kobayashi J, Ueda M, Kishimoto T, Inouye S (2009) Toward the production of quantum degenerate bosonic polar molecules, $^{41}K^{87}Rb$. New J Phys 11(5):055,035. https://doi.org/10.1088/1367-2630/11/5/055035
3. Aikawa K, Akamatsu D, Hayashi M, Oasa K, Kobayashi J, Naidon P, Kishimoto T, Ueda M, Inouye S (2010) Coherent transfer of photoassociated molecules into the rovibrational ground state. Phys Rev Lett 105(20):203,001. https://doi.org/10.1103/PhysRevLett.105.203001
4. Altaf A, Dutta S, Lorenz J, Pérez-Ríos J, Chen YP, Elliott DS (2015) Formation of ultracold $^7Li^{85}Rb$ molecules in the lowest triplet electronic state by photoassociation and their detection by ionization spectroscopy. J Chem Phys 142(11):114,310. https://doi.org/10.1063/1.4914917
5. Barbé V, Ciamei A, Pasquiou B, Reichsöllner L, Schreck F, Zuchowski PS, Hutson JM (2018) Observation of Feshbach resonances between alkali and closed-shell atoms. Nat Phys 14:881–884. https://doi.org/10.1038/s41567-018-0169-x
6. Bergmann K, Theuer H, Shore B (1998) Coherent population transfer among quantum states of atoms and molecules. Rev Mod Phys 70(3):1003. https://doi.org/10.1103/RevModPhys.70.1003
7. Berninger M, Zenesini A, Huang B, Harm W, Ngerl HC, Ferlaino F, Grimm R, Julienne PS, Hutson JM (2013) Feshbach resonances, weakly bound molecular states, and coupled-channel potentials for cesium at high magnetic fields. Phys Rev A 87(3):032,517. https://doi.org/10.1103/physreva.87.032517
8. Brue DA, Hutson JM (2012) Magnetically tunable Feshbach resonances in ultracold Li-Yb mixtures. Phys Rev Lett 108(4):043,201. https://doi.org/10.1103/PhysRevLett.108.043201
9. Brue DA, Hutson JM (2013) Prospects of forming ultracold molecules in $^2\Sigma$ states by magnetoassociation of alkali-metal atoms with Yb. Phys Rev A 87(5):052,709. https://doi.org/10.1103/physreva.87.052709
10. Cabrera CR, Tanzi L, Sanz J, Naylor B, Thomas P, Cheiney P, Tarruell L (2017) Quantum liquid droplets in a mixture of Bose-Einstein condensates. Science 359(6373):301–304. https://doi.org/10.1126/science.aao5686
11. Chin C, Vuletić V, Kerman AJ, Chu S, Tiesinga E, Leo PJ, Williams CJ (2004) Precision Feshbach spectroscopy of ultracold Cs_2. Phys Rev A 70(3):032,701. https://doi.org/10.1103/physreva.70.032701
12. Chin C, Grimm R, Julienne P, Tiesinga E (2010) Feshbach resonances in ultracold gases. Rev Mod Phys 82(2):1225. https://doi.org/10.1103/revmodphys.82.1225

---

[1]This is mostly limited by an inhomogeneous lattice light shift and for subsequent STIRAP cycles after the first cycle they observe a molecule formation efficiency of 81%.

13. Chotia A, Neyenhuis B, Moses SA, Yan B, Covey JP, Foss-Feig M, Rey AM, Jin DS, Ye J (2012) Long-lived dipolar molecules and Feshbach molecules in a 3D optical lattice. Phys Rev Lett 108(080):405. https://doi.org/10.1103/PhysRevLett.108.080405

14. Ciamei A, Bayerle A, Chen CC, Pasquiou B, Schreck F (2017a) Efficient production of long-lived ultracold sr_2 molecules. Phys Rev A 96(013):406. https://doi.org/10.1103/PhysRevA.96.013406

15. Ciamei A, Bayerle A, Pasquiou B, Schreck F (2017b) Observation of bose-enhanced photoassociation products. Europhys Lett 119(4):46,001. https://doi.org/10.1209/0295-5075/119/46001

16. Comparat D, Drag C, Tolra BL, Fioretti A, Pillet P, Crubellier A, Dulieu O, Masnou-Seeuws F (2000) Formation of cold Cs ground state molecules through photoassociation in the pure long-range state. Eur Phys J D 11(1):59–71. https://doi.org/10.1007/s100530070105

17. Cornish SL, Claussen NR, Roberts JL, Cornell EA, Wieman CE (2000) Stable ^{85}Rb Bose-Einstein condensates with widely tunable interactions. Phys Rev Lett 85(9):1795–1798. https://doi.org/10.1103/physrevlett.85.1795

18. Dalfovo F, Giorgini S, Pitaevskii LP, Stringari S (1999) Theory of Bose-Einstein condensation in trapped gases. Rev Mod Phys 71(3):463. https://doi.org/10.1103/RevModPhys.71.463

19. Danzl JG, Mark MJ, Haller E, Gustavsson M, Hart R, Aldegunde J, Hutson JM, Ngerl HC (2010) An ultracold high-density sample of rovibronic ground-state molecules in an optical lattice. Nat Phys 6(4):265–270. https://doi.org/10.1038/nphys153

20. Deiglmayr J, Grochola A, Repp M, Mörtlbauer K, Glück C, Lange J, Dulieu O, Wester R, Weidemüller M (2008) Formation of ultracold polar molecules in the rovibrational ground state. Phys Rev Lett 101(13):133,004. https://doi.org/10.1103/PhysRevLett.101.133004

21. Dowd W, Roy RJ, Shrestha RK, Petrov A, Makrides C, Kotochigova S, Gupta S (2015) Magnetic field dependent interactions in an ultracold Li-Yb($^3P_2$) mixture. New J Phys 17(5):055,007. https://doi.org/10.1088/1367-2630/17/5/055007

22. Fedichev PO, Reynolds MW, Shlyapnikov GV (1996) Three-body recombination of ultracold atoms to a weakly bound s level. Phys Rev Lett 77:2921–2924. https://doi.org/10.1103/PhysRevLett.77.2921

23. Fioretti A, Comparat D, Drag C, Amiot C, Dulieu O, Masnou-Seeuws F, Pillet P (1999) Photoassociative spectroscopy of the Cs_2 0_g^- long-range state. Eur Phys J D 5(3):389–403. https://doi.org/10.1007/s100530050271

24. González-Martínez ML, Hutson JM (2013) Magnetically tunable Feshbach resonances in Li +Yb (^3P_J). Phys Rev A 88(020):701. https://doi.org/10.1103/PhysRevA.88.020701

25. Gribakin GF, Flambaum VV (1993) Calculation of the scattering length in atomic collisions using the semiclassical approximation. Phys Rev A 48(1):546–553. https://doi.org/10.1103/PhysRevA.48.546

26. Guo M, Zhu B, Lu B, Ye X, Wang F, Vexiau R, Bouloufa-Maafa N, Quéméner G, Dulieu O, Wang D (2016) Creation of an ultracold gas of ground-state dipolar $^{23}Na^{87}Rb$ molecules. Phys Rev Lett 116(205):303. https://doi.org/10.1103/PhysRevLett.116.205303

27. Heo MS, Wang TT, Christensen CA, Rvachov TM, Cotta DA, Choi JH, Lee YR, Ketterle W (2012) Formation of ultracold fermionic NaLi Feshbach molecules. Phys Rev A 86(021):602. https://doi.org/10.1103/PhysRevA.86.021602

28. Höfer M, Riegger L, Scazza F, Hofrichter C, Fernandes DR, Parish MM, Levinsen J, Bloch I, Fölling S (2015) Observation of an orbital interaction-induced Feshbach resonance in ^{173}Yb. Phys Rev Lett 115(265):302. https://doi.org/10.1103/PhysRevLett.115.265302

29. Hutson JM (2007) Feshbach resonances in ultracold atomic and molecular collisions: threshold behaviour and suppression of poles in scattering lengths. New J Phys 9(5):152. https://doi.org/10.1088/1367-2630/9/5/152

30. Kadau H, Schmitt M, Wenzel M, Wink C, Maier T, Ferrier-Barbut I, Pfau T (2016) Observing the Rosensweig instability of a quantum ferrofluid. Nature 530:194. https://doi.org/10.1038/nature16485

31. Kato S, Sugawa S, Shibata K, Yamamoto R, Takahashi Y (2013) Control of resonant interaction between electronic ground and excited states. Phys Rev Lett 110(173):201. https://doi.org/10.1103/PhysRevLett.110.173201

32. Kerman AJ, Sage JM, Sainis S, Bergeman T, DeMille D (2004) Production of ultracold polar RbCs* molecules via photoassociation. Phys Rev Lett 92(3):033,004. https://doi.org/10.1103/physrevlett.92.033004

33. Khaykovich L, Schreck F, Ferrari G, Bourdel T, Cubizolles J, Carr LD, Castin Y, Salomon C (2002) Formation of a matter-wave bright soliton. Science 296(5571):1290–1293. https://doi.org/10.1126/science.1071021

34. Khramov A, Hansen A, Dowd W, Roy RJ, Makrides C, Petrov A, Kotochigova S, Gupta S (2014) Ultracold heteronuclear mixture of ground and excited state atoms. Phys Rev Lett 112(3):033,201. https://doi.org/10.1103/PhysRevLett.112.033201

35. Köhler T, Góral K, Julienne PS (2006) Production of cold molecules via magnetically tunable Feshbach resonances. Rev Mod Phys 78(4):1311. https://doi.org/10.1103/RevModPhys.78.1311

36. Köppinger MP, McCarron DJ, Jenkin DL, Molony PK, Cho HW, Cornish SL, Le Sueur CR, Blackley CL, Hutson JM (2014) Production of optically trapped [87]RbCs Feshbach molecules. Phys Rev A 89(3):033,604. https://doi.org/10.1103/PhysRevA.89.033604

37. Kotochigova S, Tiesinga E, Julienne PS (2000) Relativistic ab initio treatment of the second-order spin-orbit splitting of the $a^3\Sigma_u^+$ potential of rubidium and cesium dimers. Phys Rev A 63(012):517. https://doi.org/10.1103/PhysRevA.63.012517

38. Kraemer T, Mark M, Waldburger P, Danzl JG, Chin C, Engeser B, Lange AD, Pilch K, Jaakkola A, Nägerl HC, Grimm R (2006) Evidence for Efimov quantum states in an ultracold gas of caesium atoms. Nature 440(7082):315–318. https://doi.org/10.1038/nature04626

39. Lang F, Straten Pvd, Brandsttter B, Thalhammer G, Winkler K, Julienne PS, Grimm R, Hecker Denschlag J (2008) Cruising through molecular bound-state manifolds with radiofrequency. Nat Phys 4(3):223. https://doi.org/10.1038/nphys838

40. Lett PD, Julienne PS, Phillips WD (1995) Photoassociative spectroscopy of laser-cooled atoms. Annu Rev Phys Chem 46(1):423–452. https://doi.org/10.1146/annurev.pc.46.100195.002231

41. Li P, Liu W, Wu J, Ma J, Fan Q, Xiao L, Sun W, Jia S (2017) New observation and analysis of the ultracold Cs_2 0_u^+ and 1_g long-range states at the asymptote $6S_{1/2} + 6P_{1/2}$. J Quant Spectrosc Radiat Transf 196:176–181. https://doi.org/10.1016/j.jqsrt.2017.04.014

42. Lignier H, Fioretti A, Horchani R, Drag C, Bouloufa N, Allegrini M, Dulieu O, Pruvost L, Pillet P, Comparat D (2011) Deeply bound cold caesium molecules formed after 0_g^- resonant coupling. Phys Chem Chem Phys 13(42):18,910. https://doi.org/10.1039/c1cp21488h

43. Liu W, Xu R, Wu J, Yang J, Lukashov SS, Sovkov VB, Dai X, Ma J, Xiao L, Jia S (2015) Observation and deperturbation of near-dissociation ro-vibrational structure of the Cs_2 state $0_u^+ \left(a^1\Sigma_u^+ \sim b^3\Pi_u\right)$ at the asymptote $6S_{1/2}+6P_{1/2}$. J Chem Phys 143(12):124,307. https://doi.org/10.1063/1.4931646

44. Ma J, Liu W, Yang J, Wu J, Sun W, Ivanov VS, Skublov AS, Sovkov VB, Dai X, Jia S (2014) New observation and combined analysis of the Cs_2 0_g^-, 0_u^+, and 1_g states at the asymptotes $6S_{1/2}+ 6P_{1/2}$ and $6S_{1/2}+ 6P_{3/2}$. J Chem Phys 141(24):244,310. https://doi.org/10.1063/1.4904265

45. Mackie M, Kowalski R, Javanainen J (2000) Bose-stimulated Raman adiabatic passage in photoassociation. Phys Rev Lett 84:3803–3806. https://doi.org/10.1103/PhysRevLett.84.3803

46. Mark M, Ferlaino F, Knoop S, Danzl JG, Kraemer T, Chin C, Nägerl HC, Grimm R (2007) Spectroscopy of ultracold trapped cesium Feshbach molecules. Phys Rev A 76(042):514. https://doi.org/10.1103/PhysRevA.76.042514

47. McDonald M, McGuyer BH, Apfelbeck F, Lee CH, Majewska I, Moszynski R, Zelevinsky T (2016) Photodissociation of ultracold diatomic strontium molecules with quantum state control. Nature 535:122. https://doi.org/10.1038/nature18314

48. Mies FH, Tiesinga E, Julienne PS (2000) Manipulation of Feshbach resonances in ultracold atomic collisions using time-dependent magnetic fields. Phys Rev A 61(022):721. https://doi.org/10.1103/PhysRevA.61.022721

49. Moerdijk AJ, Verhaar BJ, Axelsson A (1995) Resonances in ultracold collisions of ^{6}Li, ^{7}Li, and ^{23}Na. Phys Rev A 51:4852–4861. https://doi.org/10.1103/PhysRevA.51.4852

50. Molony PK, Gregory PD, Ji Z, Lu B, Kppinger MP, Le Sueur CR, Blackley CL, Hutson JM, Cornish SL (2014) Creation of ultracold ^{87}Rb^{133}Cs molecules in the rovibrational ground state. Phys Rev Lett 113(25):255,301. https://doi.org/10.1103/PhysRevLett.113.255301

51. Molony PK, Gregory PD, Kumar A, Le Sueur CR, Hutson JM, Cornish SL (2016) Production of ultracold ^{87}Rb^{133}Cs in the absolute ground state: complete characterisation of the STIRAP transfer. ChemPhysChem 17(22):3811–3817. https://doi.org/10.1002/cphc.201600501

52. Ni KK, Ospelkaus S, de Miranda MHG, Pe'er A, Neyenhuis B, Zirbel JJ, Kotochigova S, Julienne PS, Jin DS, Ye J (2008) A high phase-space-density gas of polar molecules. Science 322(5899):231–235. https://doi.org/10.1126/science.1163861

53. Nikolov AN, Ensher JR, Eyler EE, Wang H, Stwalley WC, Gould PL (2000) Efficient production of ground-state potassium molecules at sub-mK temperatures by two-step photoassociation. Phys Rev Lett 84:246–249. https://doi.org/10.1103/PhysRevLett.84.246

54. Pagano G, Mancini M, Cappellini G, Livi L, Sias C, Catani J, Inguscio M, Fallani L (2015) Strongly interacting gas of two-electron fermions at an orbital Feshbach resonance. Phys Rev Lett 115(26):265,301. https://doi.org/10.1103/physrevlett.115.265301

55. Park JW, Will SA, Zwierlein MW (2015) Ultracold dipolar gas of fermionic ^{23}Na^{40}K molecules in their absolute ground state. Phys Rev Lett 114(205):302. https://doi.org/10.1103/PhysRevLett.114.205302

56. Petrov DS, Salomon C, Shlyapnikov GV (2004) Weakly bound dimers of fermionic atoms. Phys Rev Lett 93(090):404. https://doi.org/10.1103/PhysRevLett.93.090404

57. Regal CA, Greiner M, Jin DS (2004) Observation of resonance condensation of fermionic atom pairs. Phys Rev Lett 92(040):403. https://doi.org/10.1103/PhysRevLett.92.040403

58. Roati G, Zaccanti M, D'Errico C, Catani J, Modugno M, Simoni A, Inguscio M, Modugno G (2007) ^{39}K Bose-Einstein condensate with tunable interactions. Phys Rev Lett 99(1):010,403. https://doi.org/10.1103/physrevlett.99.010403

59. Sage JM, Sainis S, Bergeman T, DeMille D (2005) Optical production of ultracold polar molecules. Phys Rev Lett 94(20):203,001. https://doi.org/10.1103/physrevlett.94.203001

60. Schäfer F, Konishi H, Bouscal A, Yagami T, Takahashi Y (2017) Spectroscopic determination of magnetic-field-dependent interactions in an ultracold Yb($^3P_2$)-Li mixture. Phys Rev A 96(032):711. https://doi.org/10.1103/PhysRevA.96.032711

61. Stellmer S, Pasquiou B, Grimm R, Schreck F (2012) Creation of ultracold Sr$_2$ molecules in the electronic ground state. Phys Rev Lett 109(11):115,302. https://doi.org/10.1103/PhysRevLett.109.115302

62. Strecker KE, Partridge GB, Hulet RG (2003) Conversion of an atomic Fermi gas to a long-lived molecular Bose gas. Phys Rev Lett 91(080):406. https://doi.org/10.1103/PhysRevLett.91.080406

63. Taie S, Watanabe S, Ichinose T, Takahashi Y (2016) Feshbach-resonance-enhanced coherent atom-molecule conversion with ultranarrow photoassociation resonance. Phys Rev Lett 116(043):202. https://doi.org/10.1103/PhysRevLett.116.043202

64. Takasu Y, Fukushima Y, Nakamura Y, Takahashi Y (2017) Magnetoassociation of a Feshbach molecule and spin-orbit interaction between the ground and electronically excited states. Phys Rev A 96(023):602. https://doi.org/10.1103/PhysRevA.96.023602

65. Takekoshi T, Debatin M, Rameshan R, Ferlaino F, Grimm R, Nägerl HC, Le Sueur CR, Hutson JM, Julienne PS, Kotochigova S, Tiemann E (2012) Towards the production of ultracold ground-state RbCs molecules: Feshbach resonances, weakly bound states, and the coupled-channel model. Phys Rev A 85(3):032,506. https://doi.org/10.1103/PhysRevA.85.032506

66. Takekoshi T, Reichsöllner L, Schindewolf A, Hutson JM, Le Sueur CR, Dulieu O, Ferlaino F, Grimm R, Nägerl HC (2014) Ultracold dense samples of dipolar RbCs molecules in the rovibrational and hyperfine ground state. Phys Rev Lett 113(205301):205,301. https://doi.org/10.1103/physrevlett.113.205301

67. Thalhammer G, Winkler K, Lang F, Schmid S, Grimm R, Denschlag JH (2006) Long-lived Feshbach molecules in a three-dimensional optical lattice. Phys Rev Lett 96(050):402. https://doi.org/10.1103/PhysRevLett.96.050402

68. Timmermans E, Tommasini P, Hussein M, Kerman A (1999) Feshbach resonances in atomic Bose-Einstein condensates. Phys Rep 315(13):199–230. https://doi.org/10.1016/S0370-1573(99)00025-3

69. Vitanov NV, Fleischhauer M, Shore BW, Bergmann K (2001) Coherent manipulation of atoms and molecules by sequential laser pulses. Adv At Mol Opt Phy 46:55–190. https://doi.org/10.1016/S1049-250X(01)80063-X

70. Vitanov NV, Rangelov AA, Shore BW, Bergmann K (2017) Stimulated Raman adiabatic passage in physics, chemistry, and beyond. Rev Mod Phys 89(1):015,006. https://doi.org/10.1103/revmodphys.89.015006

71. Viteau M, Chotia A, Allegrini M, Bouloufa N, Dulieu O, Comparat D, Pillet P (2008) Optical pumping and vibrational cooling of molecules. Science 321(5886):232–234. https://doi.org/10.1126/science.1159496

72. Voigt AC, Taglieber M, Costa L, Aoki T, Wieser W, Hänsch TW, Dieckmann K (2009) Ultracold heteronuclear Fermi-Fermi molecules. Phys Rev Lett 102(020):405. https://doi.org/10.1103/PhysRevLett.102.020405

73. Wang F, He X, Li X, Zhu B, Chen J, Wang D (2015) Formation of ultracold NaRb Feshbach molecules. New J Phys 17(3):035,003. https://doi.org/10.1088/1367-2630/17/3/035003

74. Weber T, Herbig J, Mark M, Nägerl HC, Grimm R (2003a) Bose-Einstein condensation of cesium. Science 299(5604):232–235. https://doi.org/10.1126/science.1079699

75. Weber T, Herbig J, Mark M, Nägerl HC, Grimm R (2003b) Three-body recombination at large scattering lengths in an ultracold atomic gas. Phys Rev Lett 91(123):201. https://doi.org/10.1103/PhysRevLett.91.123201

76. Wu CH, Park JW, Ahmadi P, Will S, Zwierlein MW (2012) Ultracold fermionic Feshbach molecules of $^{23}Na^{40}K$. Phys Rev Lett 109(085):301. https://doi.org/10.1103/PhysRevLett.109.085301

77. Yamaguchi A, Uetake S, Kato S, Ito H, Takahashi Y (2010) High-resolution laser spectroscopy of a Bose-Einstein condensate using the ultranarrow magnetic quadrupole transition. New J Phys 12(10):103,001. https://doi.org/10.1088/1367-2630/12/10/103001

78. Yan M, DeSalvo BJ, Huang Y, Naidon P, Killian TC (2013) Rabi oscillations between atomic and molecular condensates driven with coherent one-color photoassociation. Phys Rev Lett 111(150):402. https://doi.org/10.1103/PhysRevLett.111.150402

79. Zabawa P, Wakim A, Haruza M, Bigelow NP (2011) Formation of ultracold $X^1\Sigma^+(v^{'}=0)$ NaCs molecules via coupled photoassociation channels. Phys Rev A 84(061):401. https://doi.org/10.1103/PhysRevA.84.061401

80. Zhang R, Cheng Y, Zhai H, Zhang P (2015) Orbital Feshbach resonance in alkali-earth atoms. Phys Rev Lett 115(135):301. https://doi.org/10.1103/PhysRevLett.115.135301

81. Zirbel JJ, Ni KK, Ospelkaus S, Nicholson TL, Olsen ML, Julienne PS, Wieman CE, Ye J, Jin DS (2008) Heteronuclear molecules in an optical dipole trap. Phys Rev A 78(013):416. https://doi.org/10.1103/PhysRevA.78.013416

82. Zuchowski PS, Aldegunde J, Hutson JM (2010) Ultracold RbSr molecules can be formed by magnetoassociation. Phys Rev Lett 105(15):153,201. https://doi.org/10.1103/physrevlett.105.153201

83. Zwierlein MW, Stan CA, Schunck CH, Raupach SMF, Kerman AJ, Ketterle W (2004) Condensation of pairs of fermionic atoms near a Feshbach resonance. Phys Rev Lett 92(120):403. https://doi.org/10.1103/PhysRevLett.92.120403

Chapter 3
Experimental Setup

The experimental apparatus used to cool and trap Cs and Yb atoms is described in this chapter. The design of the vacuum system, magnetic field coils and the majority of the laser systems are reported in previous publications [15, 17, 21] and theses on the experiment [6, 11, 20]. Therefore, in this chapter we briefly review the essential components of the experimental apparatus and comment on their application in subsequent experiments.

3.1 Atomic Properties

3.1.1 Caesium

Caesium (Cs) is the heaviest stable element of the alkali metals and has only one stable isotope, ^{133}Cs. It is solid at room temperature but has a melting point of only 28 °C. ^{133}Cs has a single valence electron and a nuclear spin of $I = 7/2$ which make it a composite boson. The electronic structure of Cs has been extensively studied and the hyperfine splitting of the $6^2 S_{1/2}$ ground state is used to define the SI second [27]. The relevant atomic transitions for laser cooling of Cs are shown in Fig. 3.1a.

Cs has a favourable energy level structure for laser cooling. The $6^2 S_{1/2} \rightarrow 6^2 P_{3/2}$ transition at 852 nm, referred to as the D_2 line, possesses a closed transition on the $F = 4 \rightarrow F' = 5$ transition highlighted in Fig. 3.1a. Also important is the repump transition $F = 3 \rightarrow F' = 4$ which is required to close off the cooling cycle by preventing atoms decaying into the $F = 3$ ground state. Here, we use F and F' to denote the hyperfine quantum number of the ground and excited state respectively.

© Springer Nature Switzerland AG 2019

A. Guttridge, *Photoassociation of Ultracold CsYb Molecules and Determination of Interspecies Scattering Lengths*, Springer Theses, https://doi.org/10.1007/978-3-030-21201-8_3

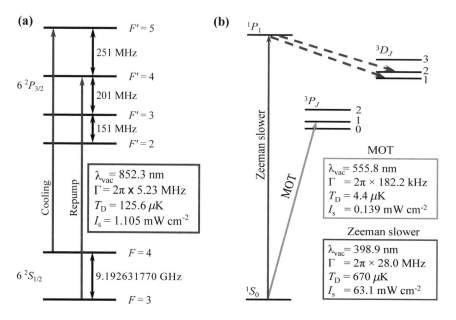

Fig. 3.1 **a** Relevant atomic transitions for laser cooling of caesium. **b** Energy level diagram of ytterbium showing the dominant 399 nm singlet transition to the $^1P_1$ state and the narrow 556 nm intercombination transition to the $^3P_1$ state

3.1.2   Ytterbium

Ytterbium (Yb) is a lanthanide element with two valence electrons. Yb possesses seven stable isotopes, five bosons and two fermions. The scattering lengths of these isotopes cover a broad range from positive to vanishingly small to large and negative. This demonstrated ability to tune the scattering length using the reduced mass is important for the investigation of Cs+Yb mixtures [5]. The scattering lengths of all the stable Yb isotopes are presented in Table 3.1 along with their relative abundance and nuclear spin. All isotopes, except for ^{172}Yb which possesses a large negative scattering length, were first cooled to quantum degeneracy by the Kyoto group [12, 13, 36–39].

The relevant electronic structure of Yb is shown in Fig. 3.1b. Due to ytterbium's two valence electrons in the $6s$ orbital, it's electronic structure is remarkably similar to alkaline-earth elements like strontium and calcium, whereby the electronic states are separated into singlet ($S = 0$) and triplet ($S = 1$) manifolds. Electronic spin changing transitions $\Delta S = 1$ are forbidden by the electric dipole selection rules, which would prohibit excitation of the Yb atoms from the ground state ($S = 0$) to the $^3P_1$ state ($S = 1$). However, in heavy atoms $j - j$ spin-orbit coupling introduces a small admixture of the $^1P_1$ state into the $^3P_1$ state, causing the $^1S_0 \rightarrow ^3P_1$ transition to be weakly allowed.

Table 3.1 All stable isotopes of Yb with their respective natural abundance, nuclear spin and intraspecies scattering length

Yb isotope	Abundance [9] (%)	Nuclear spin	Scattering length (a_0) [23]
^{176}Yb	12.70	0	−24
^{174}Yb	31.80	0	105
^{173}Yb	16.10	5/2	199
^{172}Yb	21.90	0	−599
^{171}Yb	14.30	1/2	−3
^{170}Yb	3.05	0	64
^{168}Yb	0.13	0	252

In Yb, laser cooling can be accomplished on both the 399 nm $^1S_0 \rightarrow{}^1 P_1$ transition and the 556 nm $^1S_0 \rightarrow{}^1 P_1$ transition. The $^1D_2$ state lies above the $^1P_1$ state, which means that the large decay path that occurs in strontium is absent in ytterbium. However, the $^1P_1$ state may still decay to the triplet $^3D_{1,2}$ states which limits the total atom number achieved in a MOT without repumping light [16]. The closed, two-level transition at 556 nm is more suitable for the MOT. This transition has a narrow linewidth of 182 kHz which corresponds to a low Doppler temperature of 4.4 μK. However, the narrow linewidth results in a greatly reduced MOT capture velocity of 7 m/s; this requires precise operation of the Zeeman slower. Therefore, we use the strong 399 nm transition for Zeeman slowing as the large linewidth and short wavelength lead to a large maximum deceleration.

3.2 Experimental Overview

The main components of our setup for trapping and cooling of Cs and Yb are presented in this section. The majority of the optical and vacuum systems are distributed over two optical tables. One optical table, referred to as the 'laser table', is used for the Cs and Yb laser systems. The only exception being the Yb 399 nm laser which is mounted on a breadboard above the main experimental table. On the laser table, the frequencies of the Cs and Yb lasers are stabilised using atomic spectroscopy performed in a vapour cell for Cs and in an atomic beam for Yb. An array of acousto-optic modulators (AOMs) are used to generate the different frequencies of light required for both species. The light required in the experiment is transported over to the main experimental table using polarisation-maintaining single-mode optical fibres. The intensity of light transported to the main experiment is controlled using AOMs and optical shutters, both of which are controlled by the experimental control software.

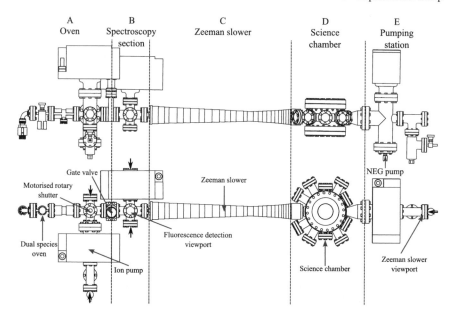

Fig. 3.2 Schematic drawings of the vacuum system from a bird's eye and side view. The dashed lines split the apparatus into the labelled sections which are described in the text: A the dual species oven; B the spectroscopy cross; C the Zeeman slower; D the science chamber; and E the pumping station that maintains UHV pressures in the science chamber

Aside from the laser table, the other optical table is labelled as the 'main experimental table'. This table contains the whole vacuum apparatus, dipole trapping lasers and optical fibres for cooling and trapping light.

Experiments with ultracold atoms require ultrahigh vacuum (UHV) to reduce the number of collisions with background gas in the chamber. Collisions with the room temperature background gas results in tremendous heating of the sample and cause atom loss. However, for trapping a large number of atoms we also require a large enough vapour pressure of both species for sufficient atom flux in the trapping region. This may lead to partial pressures outside the UHV regime. To resolve this issue we divide our vacuum system into multiple sections as shown in Fig. 3.2, which allows control of the pressure in different sections using differential pumping.

The first section of the vacuum system is the dual-species oven, which produces the necessary atomic beams of Cs and Yb. Following this section is a six-way cross for probing of the atomic beams in flux measurements. A dual-species Zeeman slower is used to slow the Yb and Cs atomic beams and allows Cs and Yb MOTs to be loaded in the stainless steel science chamber.

3.2.1 Dual-Species Oven

The major difficulty in the design of the dual-species oven is the vast difference between the room temperature vapour pressures of Cs and Yb. For Yb the vapour pressure is 3×10^{-21} torr [1] at room temperature, therefore, an effusive oven is required to create a high-flux, collimated atomic beam suitable for slowing and trapping. To reach a suitable vapour pressure (10^{-3} torr) the Yb source must be heated to over 400 °C. On the other hand, Cs has a vapour pressure of 7×10^{-7} torr at room temperature [35] which is sufficient to load a MOT directly from the vapour. However, in the interest of maintaining UHV conditions in the MOT region a collection MOT or Zeeman slowed atomic beam is desirable. Therefore, we utilise an atomic beam of Cs and Yb which is decelerated by a Zeeman slower before the MOT.

A schematic of our dual-species oven is shown in Fig. 3.3. Two separate sections are used to house the Cs and Yb sources. For the Cs source we use an ampoule containing 1 g of Cs, which is placed in the bellows at the left hand end of Fig. 3.3a. After the oven section has been evacuated and baked, the ampoule is broken to release the Cs. Verifying the ampoule is sufficiently broken is a difficult task. Initially, we bent the bellows (quite forcefully) until we heard a crunching sound of the glass ampoule breaking. However, years later when recharging the Cs in the oven (which we assumed had run out) we discovered that we still had an almost full ampoule of Cs. The crunching glass sound had been the top of the ampoule breaking, which does not contain any Cs as the Cs resides in a reservoir in the lower half. This lower half had a small crack from which the Cs had been leaking out but this crack had been sealed over by some condensed substance, preventing the flow of Cs atoms into the chamber. Now when recharging the oven, we remove most of the upper part of the ampoule and carefully score the glass of the lower half (containing the Cs) to ensure that the lower half of the ampoule breaks sufficiently.

The Yb section of the oven is loaded using a 5 g ingot of the element. The Cs and Yb sections of the oven are separated by a valve which allows the Cs oven to be sealed off when not in use. The Cs oven section is typically heated to 85 °C and the Yb part to 485 °C. The atomic beams emanating from the sources travel through two separate channels of a semicircular cross section and at the end are incident on the capillaries. The capillaries are used to collimate the atomic beams and are composed of 55 capillary tubes in a triangular array. The housing of the capillary tubes is shown in Fig. 3.3c.

The temperature gradients provided by four band heaters are shown in Fig. 3.3a. The gradients are required to ensure the sources of Cs and Yb do not migrate from their reservoirs. To prevent blocking of the capillary tubes, the capillary section of the oven is heated to the highest temperatures. The end of the oven is connected to a six-way cross, which is followed by a gate valve. The gate valve allows the oven to be serviced without letting the Zeeman slower and science chamber up to air. Also connected to the six-way cross is a rotary feedthrough which is used to rotate an atomic beam shutter.

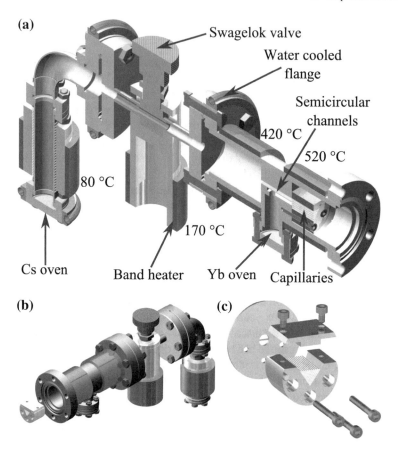

Fig. 3.3 **a** A cutaway schematic of the dual species oven showing the paths of the two species through the oven. The nozzle heaters are shown as copper bands and the water cooling as copper tubes. The temperatures displayed are the nozzle heater set points, the Yb chamber is typically measured to be at 485 °C. **b** Rendering of the oven with the capillary clamp shown outside the oven for detail. **c** Exploded view of the capillary array and the wire-eroded semicircular channels that the array is clamped over within the oven

3.2.2   Zeeman Slower

The dual-species Zeeman slower is shown in Section C of Fig. 3.2. The design and construction of the Zeeman slower is extensively discussed in Ref. [17] and in previous theses on the experiment [6, 11, 20]. To preserve optical access in the science chamber we chose to use a single Zeeman slower for both species. This requires only a single line of entry for the atoms and one viewport for the slowing light. The contrasting properties of Cs and Yb required a flexible design that can optimally slow both species. Yb imposes the most stringent requirements as loading a MOT on the $^1S_0 \rightarrow ^3P_1$ transition requires the atoms entering the capture region of the MOT to be travelling slower than 7 m/s. To prevent the slowly moving atoms from falling too much under gravity as they travel to the MOT region, the Zeeman slower exit must be as close as possible to the MOT. The Zeeman slower exit is located 7.5 cm from the MOT, a sufficient distance to allow the large magnetic field at the end of the slower to drop to near zero. The main novelty of the slower design is the coil windings that allow tuning of the field profile for slowing of Cs or Yb. The coil windings used are discussed in Sect. 3.4.1.

3.2.3 Science Chamber

The focal point of the experimental apparatus is the science chamber, labelled as section D in Fig. 3.2. A pumping station (section E) is attached to the science chamber to maintain UHV pressures. It is within the science chamber where all of the experiments detailed in this thesis take place.

   A large amount of optical access is required for any dual-species experiment, the science chamber has ten anti-reflection (AR) coated viewports oriented horizontally and another two vacuum ports for the Zeeman slower light and the pumping station. The vast array of distinct laser beams currently used in the setup is shown in Fig. 3.4.

   It is convenient to introduce a coordinate system to describe the orientation of beams that will be used throughout this thesis. The East-West axis is along the Zeeman slower axis, with East defined to be the direction atoms travel from the oven to the

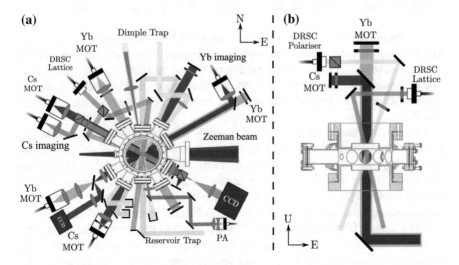

Fig. 3.4 Schematic of beams and optics around the science chamber. **a** Beams and optics in the horizontal plane. **b** Beams and optics in the vertical plane. For clarity the MOT and Bias coils are not shown

chamber. The North-South axis is also in the horizontal plane but perpendicular to the Zeeman slower, along the dimple trap and photoassociation viewport. The vertical direction is simply referred to as Up and Down.

The majority of the beams intersect the viewports of the science chamber in the horizontal plane shown in Fig. 3.4a. We use a standard six-beam configuration for both Cs and Yb MOTs. For Cs, fibre cage systems are used to produce collimated MOT beams with a $1/e^2$ diameter of 18.0(1) mm, whereas the Yb fibre cage systems produce MOT beams of diameter 24.5(2) mm. Quarter waveplates are placed after the cage systems and before the retro-reflecting mirrors to create circularly polarised light of the correct handedness. The laser setup for the generation of the Yb 556 nm MOT light is described in Sect. 3.3.2 and the Cs setup in Sect. 3.3.1.

The Yb imaging is performed on the 399 nm $^1S_0 \rightarrow{}^1 P_1$ transition using a beam with a $1/e^2$ diameter of 13.6(2) mm. The 399 nm probe light is overlapped with the 556 nm MOT light using a dichroic mirror and separated after the atoms using another dichroic mirror. This dichroic mirror is used to align the probe onto the CCD camera (Andor Luca) used for Yb absorption imaging. For absorption imaging of Cs, the MOT and probe beams are overlapped on a polarising beam splitter (PBS) cube. At the opposite end of the chamber, the beams are separated using another PBS and the probe beam is incident on a CCD camera (Andor Ixon). More information on the absorption imaging setup is presented later in Sect. 3.5.

The Zeeman slower beams for both species are focused using independent systems of lenses before the two Zeeman slower beams are overlapped. The Cs Zeeman slower beam is overlapped with the Yb 399 nm light on a dichroic mirror and both beams are focused to a position near the oven capillaries, $\simeq$2 m away form the Zeeman slower viewport. More information on the laser setup for the 399 nm light is presented in Sect. 3.3.2. The generation of all the Cs light used in the experiment is explained in Sect. 3.3.1.

The DRSC lattice beams are split into two paths by a PBS after the fibre output. The running wave beam passes through the Cs imaging viewport and the horizontal standing wave beam passes through the central dimple trap viewport. On the opposite side of the chamber the horizontal standing wave beam is retro-reflected to create a standing wave. More details on the implementation of DRSC in the setup is given in Sect. 5.2.

Two distinct optical dipole traps are currently used in the setup: the dimple trap and the reservoir trap. The dimple trap light is generated by a 100 W IPG and is used for trapping both Cs and Yb. The two arms of the trap are focussed through different viewports, dimple beam 1 through the central dimple viewport and dimple beam 2 through the Cs MOT viewport. The setup of the dimple trap is detailed in Sect. 4.1.1 and its implementation in the production of Cs BEC is described in Sect. 5.3.2.

The reservoir trap light is generated by a 50 W IPG laser and is solely used for trapping Cs. We use a bow-tie geometry for the trap, with the ingoing beam passing through the central dimple viewport and the returning beam passing through the Cs MOT viewport. The setup of this trap is described in Sect. 5.3.1.

Finally, the photoassociation (PA) light is beam shaped using a fibre cage system and then counterpropagated against dimple beam 1 using a dichroic mirror. The laser systems used to produce the photoassociation light are described in Sect. 7.2.

It is not immediately apparent why so many of the beams pass through the central dimple viewport (and Cs MOT viewports) as opposed to the Yb MOT viewports which have fewer beams. This is because of the AR coatings of the viewports, the Yb viewports have a reflectance of $\simeq$10% per face at 1064 nm, so all infrared beams (which is the majority) must pass through the Cs MOT or central viewports. It will therefore be very challenging to install a 3D (infrared) lattice in the current setup.

Figure 3.4b shows the optical setup in the vertical plane which is far less congested than the horizontal plane. The upper and lower viewports of the science chamber are also AR coated, but a broader AR coating is used than the Yb viewports. The reflectance at 1064 nm is around ~4% per face. The Yb and Cs MOT beams are overlapped on a dichroic mirror below the science chamber and then split apart using a dichroic above the chamber. Also along the vertical axis are two additional DRSC beams: the vertical running lattice beam and the polariser beam.

The cut-through of the science chamber in Fig. 3.4b shows the re-entrant flanges incorporated on the top and bottom of the science chamber. These allow multiple sets of coils to be mounted as close to the atoms as possible. This ensures homogeneity of magnetic fields applied by the coils and is especially important when applying the large magnetic fields of up to ~2000 G, for which the coils have been made. Further information on the magnetic field coils used in the experiment may be found in Sect. 3.4.2.

3.3 Laser Systems for Trapping and Cooling Cs and Yb

The number of different frequencies of light required for cooling and trapping both Cs and Yb is illustrated in Fig. 3.4. In this section we outline the lasers and optical components used to generate all these frequencies of light.

3.3.1 Cs Laser Systems

The 852 nm light used for decelerating, cooling and imaging of Cs is provided by three lasers. For transitions from the $6S_{1/2}$, $F = 4$ state of Cs the light is provided by a Toptica DL 100 Pro diode laser. The frequency of this laser is stabilised to a detuning of -385.4 MHz from the $F = 4 \rightarrow F' = 5$ transition. A small fraction of the laser output is double-passed through an AOM which generates the required frequency difference to atomic resonance. After this AOM (AOM 2), the laser light passes through an electro-optic modulator (EOM) and a Cs vapour cell where modulation transfer spectroscopy [29, 34] is used to generate an error signal for stabilisation of the laser frequency. Another small fraction of the laser output is double passed through AOM 1 and then coupled into an optical fibre for absorption imaging of the Cs atoms. The remaining fraction of the laser output is double passed through AOM 3 and sent via an optical fibre to seed a tapered amplifier (Toptica BoosTA). The large output power of the BoosTA is required to generate the Zeeman slower cooling light, MOT cooling light and DRSC lattice light. All of these beams are required to operate at different frequencies, this is achieved using a series of different AOMs. The laser detunings from relevant transitions are illustrated in Fig. 3.5 and tabulated in Table 3.2.

The main cooling transition of the Cs MOT operates on the 'closed' $F = 4 \rightarrow F' = 5$ transition. However, MOT light detuned from the $F = 4 \rightarrow F' = 5$ transition has a non-zero probability to off-resonantly excite an atom to the $F' = 4$ excited level which may decay to the $F = 3$ ground state. The $F = 3$ state is dark to the cooling

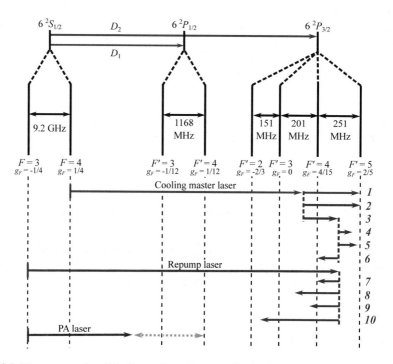

Fig. 3.5 The upper section of the figure shows the energy levels of the D_1 and D_2 lines of Cs. The hyperfine states associated with each line are shown, with labels denoting the value of F and g_F factors [35]. The lower section of the figure shows schematically the approximate frequencies of the various Cs lasers and beams used in the experiment. The shorter arrows denote the frequency shifts given by AOMs and the red numbers label the AOM used to generate the required frequency: 1 is the imaging probe, 2 is the cooling spectroscopy. 3 is the BoosTA seed, 4 is the Zeeman slower cooling beam, 5 is the MOT cooling light, 6 is the DRSC lattice beams, 7 is the repump spectroscopy, 8 is the Zeeman slower repump light, 9 is the MOT repump light and 10 is the DRSC polariser light. The Cs photoassociation laser(s) (discussed in Chaps. 7 and 8) are tunable to a broad range of frequencies red-detuned from the D_1 line

Table 3.2   Frequency detunings of Cs laser beams used in the experiment

	Transition	Detuning (MHz)
Cooling spectroscopy	$F = 4 \rightarrow F' = 5$	0.0
Cooling master laser	$F = 4 \rightarrow F' = 5$	-385.4
Imaging probe	$F = 4 \rightarrow F' = 5$	$+1.2$
BoosTA	$F = 4 \rightarrow F' = 5$	-132.0
Zeeman slower cooling	$F = 4 \rightarrow F' = 5$	-49.4
MOT cooling	$F = 4 \rightarrow F' = 5$	-9.0
Repump spectroscopy	$F = 3 \rightarrow F' = 4$	0.0
Repump master laser	$F = 3 \rightarrow F' = 4$	$+80.0$
Zeeman slower repump	$F = 3 \rightarrow F' = 4$	-32.5
MOT repump	$F = 3 \rightarrow F' = 4$	-5.4
Optical pumping	$F = 3 \rightarrow F' = 2$	$+10$

light and the atom is lost from the main cooling cycle. To alleviate this potential loss mechanism we also use repump light operating on the $F = 3 \rightarrow F' = 4$ transition which repumps the atoms back into the main cooling cycle. A Toptica DL Pro diode laser is used to generate light for driving transitions from the $6S_{1/2}$, $F = 3$ state of Cs. The output frequency of the laser is detuned by $+80$ MHz from the $F = 3 \rightarrow F' = 4$ transition using a similar scheme to the main cooling laser. The light is single-passed through AOM 7 to generate the appropriate laser detuning and the laser frequency is stabilised using frequency modulation spectroscopy [2, 14]. The remaining laser power is split between Zeeman slower repump, MOT repump and DRSC polariser fibres. The frequencies of these beams are controlled via independent AOMs like in the main cooling setup. The 0th order of the ZS repump AOM (AOM 8) is used to stabilise the length of the optical cavity used in photoassociation measurements (see Chap. 7).

A schematic of the whole Cs optical setup is shown in Fig. 3.6. The general optical layout is largely unchanged from the setup reported in previous theses on the experiment [6, 11, 20]. The most significant changes are the implementation of DRSC in the experiment and the setup of the cavity stabilisation used for photoassociation. The implementation of DRSC is discussed in Chap. 5 and the photoassociation setup is described in Chap. 7.

3.3.2   Yb Laser Systems

The complexity of the Cs laser setup is in stark contrast to the Yb laser setup. For trapping and cooling of the Yb MOT we require two frequencies of light that are provided by two lasers. The 556 nm laser generates the light for the MOT and a 399 nm laser produces the light for Zeeman slowing and absorption imaging.

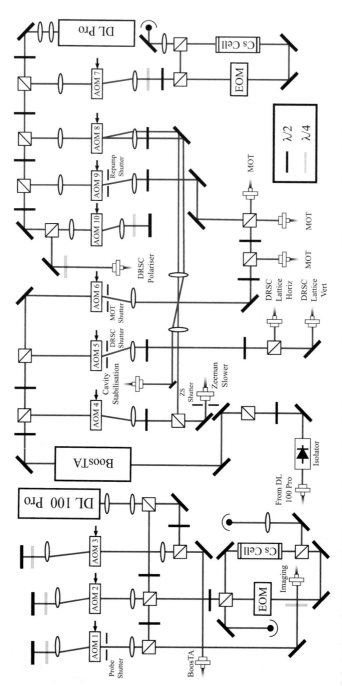

Fig. 3.6 A schematic drawing of the Cs optical setup. Cooling light is produced close to the $F = 4 \rightarrow F' = 5$ transition by the Toptica DL 100 Pro and Toptica BoosTA on the left of the figure, whereas repump light near the $F = 3 \rightarrow F' = 4$ transition is produced by the Toptica DL Pro on the right of the figure

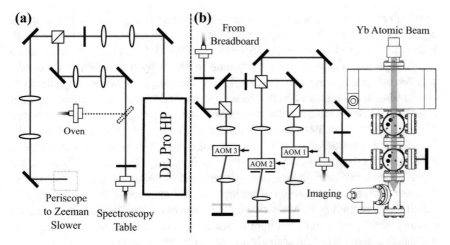

Fig. 3.7 Optical layout of the 399 nm laser system. **a** Optical setup mounted on a breadboard above the Zeeman slower viewport (right hand end of Fig. 3.2). The majority of the light is sent down a periscope for Zeeman slowing and the remainder of the light is transferred to the laser table via an optical fibre. **b** 399 nm setup on the laser table. Light from the laser breadboard is double passed through an array of AOMs to generate the necessary Zeeman slower and probe light detunings. Part of the light is split off to be used as probe light for absorption imaging. The remaining light is used for fluorescence spectroscopy of an Yb atomic beam

399 nm Light

The optical layout of the 399 nm light used in the experiment is shown in Fig. 3.7. The optics for the 399 nm light are in two distinct locations. Figure 3.7a shows the setup on a breadboard mounted above the main experimental table. A Toptica DL Pro HP diode laser is used to generate 100 mW of 399 nm light. The majority of the laser output is shaped using a telescope and sent down to the main experimental table using a periscope. The poor fibre coupling efficiency of 399 nm light necessitates the use of a periscope to transfer the most power from the laser breadboard into the vacuum chamber. After the periscope, the beam passes through a beam shutter and several lenses before the Zeeman slower viewport. This sequence of lenses forms a waist of 307(8) μm at a distance of 1.93(1) m from the Zeeman slower viewport, very close to the capillaries in the oven section. The remainder of the 399 nm laser output is coupled into an optical fibre which transports the light to the laser table where the detuning with respect to atomic resonance is set using a series of AOMs. A flipper mirror is also installed on the breadboard. When the mirror is up, light is coupled into an optical fibre to the oven section of the experiment. The output of this fibre is used for spectroscopy of the atomic beam effusing from the dual-species oven in the main vacuum system.

Figure 3.7b shows the optical layout of the 399 nm optics on the laser table. The large $\simeq -600$ MHz Zeeman slower detuning is provided by a series of double passed AOMs. All of the light from the breadboard fibre is double-passed through a 200 MHz

AOM (AOM 3), before being split into light for spectroscopy (AOM 1) and imaging (AOM 2). Both spectroscopy and imaging beams are double passed through 100 MHz AOMs. This gives the required $\simeq -600$ MHz detuning of the Zeeman slower light and allows the probe detuning to be set by the frequency difference between the spectroscopy and imaging AOMs. The imaging beam is coupled into an optical fibre and transported to the main experimental table. The spectroscopy light is sent to the "Beam Machine" for fluorescence spectroscopy of the Yb atomic beam. The frequency of the 399 nm laser is stabilised to an error signal derived from the fluorescence spectroscopy signal. The current of the 399 nm laser is modulated at 3.7 kHz, with a depth of ± 0.5 MHz on the laser frequency. The resulting fluorescence signal is demodulated by a lock-in amplifier to generate the dispersive error signal. An issue with the current setup of multiple AOMs to generate the ZS detuning, is the difficulty in finding AOMs with good AR coatings at this wavelength. Currently a large ($>80\%$) fraction of the light is lost after being double passed through two AOMs. This issue is not satisfyingly solved in the current setup.

556 nm Light

The laser setup for the 556 nm light is even simpler than the 399 nm setup. The optical layout of the 556 nm setup is shown in Fig. 3.8. The 556 nm light is generated by the now discontinued MenloSystems 'Orange One'. The system is composed of an amplified fibre laser operating at 1111.6 nm which is frequency doubled by a second-harmonic generation (SHG) module from NTT photonics. The frequency of the laser maybe be broadly tuned using the temperature of the seed laser or more finely tuned using a piezo element in the seed laser substrate. Unfortunately, we have experienced multiple failures of the SHG module of this laser system, which each time required waiting over three months for a replacement module from Japan. The cause of these issues seemed to be a combination of a high piezo slew rate and the factory set output power of the fibre laser being greater than the damage threshold of the SHG module. As a result of these multiple failures, Menlo has reduced the output power of the fibre laser to a maximum of 0.5 W and installed a shutdown feature to prevent damage to the SHG module from so-called 'monster pulses' generated by a high piezo slew rate. As a result of these changes the maximum output power of the SHG module is now 200 mW.

The output of the SHG module is split into two by a PBS cube after the laser output. A small fraction of the light is used for stabilising the laser to atomic resonance. This spectroscopy light is single-passed through the 200 MHz spectroscopy AOM (AOM 1), the AOM is used to provide a small modulation (dither) to the frequency of the light to allow frequency stabilisation to the atomic fluorescence. The light after AOM 1 is expanded by a telescope to 1.95 mm, in order to increase the interaction volume of the laser light with the atomic beam. AOM 1 applies a dither at a frequency of 3 kHz to the AOM shift centred at 217 MHz that results in optical frequency excursions of depth ± 2 MHz. The detected fluorescence from the atomic beam is demodulated using a lock-in amplifier yielding a dispersive error signal. Care must be taken that the spectroscopy beam intersects the atomic beam at exactly 90° as small deviations from orthogonality can lead to a Doppler shift of resonance resulting in

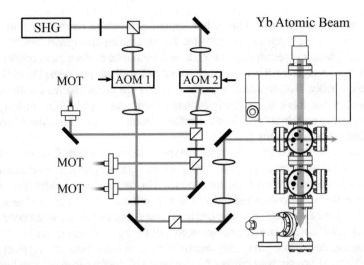

Fig. 3.8 Optical layout of the 556 nm laser system. The output of the SHG crystal is split into two branches, one for the MOT beams and the other for spectroscopy. The MOT light is single-passed through an AOM and split into three beams. Each beam is coupled into a optical fibre and sent to the main table. The spectroscopy light passes through an AOM and then intersects the Yb atomic beam of the beam machine at 90°

a systematic offset in the lock. A shift of $\simeq 0.01°$ will result in a 1 MHz frequency offset, which causes significant changes to a Yb MOT operated at low MOT light intensity. Any offset in the lock can be verified by spectroscopy of the magnetically insensitive $m_J = 0 \rightarrow m'_J = 0$ transition [15].

The majority of the output from the SHG is transmitted through the PBS and is used for the MOT beams. This light is single-passed through the 200 MHz MOT AOM (AOM 2) and split into three beams, with each beam coupled into a separate fibre to the main experiment table. Intensity and detuning ramps of the MOT light are achieved using AOM 2. The detuning of the MOT light is set by the frequency difference between AOM 1 and AOM 2.

Ytterbium Fluorescence Spectroscopy

Fluorescence spectroscopy of an Yb atomic beam is used to stabilise both 399 nm and 556 nm lasers. A comprehensive report on our Yb fluorescence spectroscopy setup is presented in Ref. [15] and in the thesis of Kemp [20]. Here I will briefly summarise the setup and how we perform fluorescence spectroscopy. This information is repeated here because the performance of the spectroscopy is linked to the current capability of the experiment to trap and cool different Yb isotopes.

Generating a suitable locking signal from atomic spectroscopy is slightly more complex in Yb compared to Cs. Spectroscopy of Cs is usually performed in glass vapour cells at room temperature. This is because at room temperature the vapour pressure is high enough to achieve a large amount of absorption of resonant light, which allows production of spectroscopy signals with a high signal-to-noise ratio

(SNR). In contrast to Cs, Yb has a low vapour pressure at room temperature [7]. When heated to the temperatures required for significant absorption ($\geq$420 °C) Yb reacts with glass, so spectroscopy of Yb atoms cannot be performed in conventional vapour cells. Designs have been created that circumvent this problem [18, 19] but are necessarily bulky. Hollow-cathode lamps [22, 25] are an alternative for absorption spectroscopy but these significantly broaden the features ($\geq$1 GHz), making sub-MHz frequency stabilisation difficult. Another alternative is stabilisation of lasers to a high finesse optical cavity [28, 41].

We employ a method using fluorescence spectroscopy of a high-flux atomic beam, similar to some other groups [26, 31]. We utilise a separate portable oven, named the 'Beam machine', not the dual-species oven used in the main experiment. This allows the atomic flux to be controlled independently to the main system and the smaller size allows the beam machine to be easily removed and recharged when necessary.

Figure 3.9 illustrates the main components of the beam machine used to conduct fluorescence spectroscopy. It is composed of a steel vacuum chamber with an external oven heater and two six-way crosses for excitation of fluorescence by laser light and subsequent fluorescence detection. Similar to the main oven, an array of parallel capillary tubes are used to produce a bright, collimated atomic beam. The band heater surrounding the oven section is typically heated to over 450 °C. A horizontal, collimated Yb atomic beam then effuses from the capillaries and passes through two further apertures, the first of diameter 6 mm, the second of diameter 8 mm.

A horizontal laser beam at either 556 or 399 nm intersects the atomic beam at 90°. The resulting resonance fluorescence is collected by custom built detectors based on designs presented in Refs. [3, 33]. The high gain of the detector allows us to run

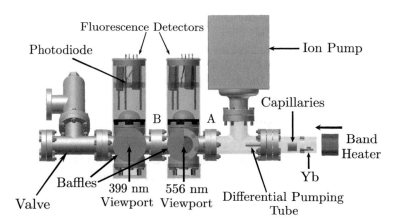

Fig. 3.9 Overview of the Yb atomic beam spectroscopy apparatus. An Yb oven provides an atomic beam (right to left on figure) which is collimated by an array of capillary tubes and passes through a differential pumping tube and two further circular apertures denoted A and B (6 and 8 mm). Optical access for the laser beams is provided separately through the horizontal viewports of two six-way crosses, with the atomic fluorescence detected in the vertical direction by two photodiode assemblies

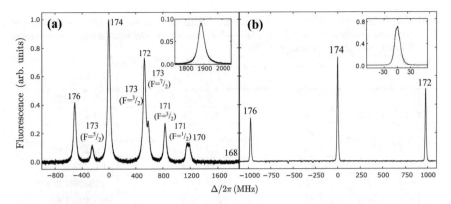

Fig. 3.10 a Typical fluorescence signal obtained by scanning the 399 nm laser with intensity $I =$ 0.1 I_{sat} and heater temperature $T = 470$ °C. The inset shows a scan over the transition in ^{168}Yb at $T = 540$ °C in black and a Voigt fit to the data in red. **b** Fluorescence signal obtained by scanning the 556 nm laser with intensity $I = 50$ I_{sat} and temperature $T = 470$ °C. In the 556 nm spectrum, the other isotopes lie outside the spectral range shown in the figure. The inset shows an enhanced view of the ^{174}Yb resonance in black and a Voigt fit to the data in red

the atomic beam at relatively low flux, thereby conserving atoms and extending the lifetime of the Yb source. At our oven operating temperature of 470 °C, measured at the external band heater, we have so far found that 5 g of Yb has lasted for over four years with usage of 10 h/day.

Examples of fluorescence spectra obtained by scanning the 399 nm and the 556 nm lasers are shown in Fig. 3.10. In the spectroscopy of the $^1S_0 \rightarrow ^1 P_1$ transition in Fig. 3.10a, we observe peaks corresponding to all the isotopes of ytterbium with the exception of ^{168}Yb, which has an extremely low natural abundance of 0.13% [9]. However, in the inset, we clearly see ^{168}Yb with a SNR greater than 100 by increasing the oven temperature to 540 °C. Table 3.3 shows the natural abundances of all Yb isotopes in addition to the isotope shifts for the 399 nm and 566 nm transitions.

The transverse velocity distribution of the atomic beam leads to Doppler broadening of the observed line shapes. The effect of Doppler broadening on the spectra at 399 nm creates a line shape described by a Voigt profile with FWHM of 40.0(2) MHz. The Gaussian contribution to the Voigt profile is dominated by 19.5(1) MHz of Doppler broadening, while the Lorentzian contribution is 30.0(1) MHz. Figure 3.10b shows spectroscopy of the narrower $^1S_0 \rightarrow ^3 P_1$ transition. The inset of the figure shows the ^{174}Yb peak fitted with a Voigt profile, the FWHM extracted from the fit is 15.0(2) MHz. As the linewidth of the 556 nm transition is significantly narrower than the 399 nm transition, the dominant broadening mechanism is Doppler broadening due to the transverse velocity of the atomic beam.

The spectroscopy features for ^{174}Yb are well resolved for both transitions. Frequency stabilisation of both 556 and 399 nm lasers performs well when stabilised to the ^{174}Yb peak, as evidenced by later measurements in Sect. 4.1.2. However, when working with different Yb isotopes, the performance of the stabilisation is not as

Table 3.3 A table showing the natural abundance for the seven stable isotopes of ytterbium and the isotopes shifts of the $^1S_0 \rightarrow ^3P_1$ and $^1S_0 \rightarrow ^1P_1$ transitions. The isotope shifts are presented relative to the most abundant isotope, ^{174}Yb

Yb isotope	Abundance [9] (%)	Shift from ^{174}Yb (MHz)	
		$^1S_0 \rightarrow ^1P_1$ [24]	$^1S_0 \rightarrow ^3P_1$ [32]
^{176}Yb	12.70	−508.89(9)	−954.832(60)
^{173}Yb ($F' = 5/2$)	16.10	−250.78(33)	2311.411(85)
^{174}Yb	31.80	0	0
^{173}Yb ($F' = 3/2$)	16.10	515.975(200)	3807.278(134)
^{172}Yb	21.90	531.11(9)	1000.020(85)
^{173}Yb ($F' = 7/2$)	16.12	589.75(24)	−2386.704(85)
^{171}Yb ($F' = 3/2$)	14.30	835.19(20)	3804.608(100)
^{171}Yb ($F' = 1/2$)	14.30	1153.68(25)	−2132.063(85)
^{170}Yb	3.05	1190.36(49)	2286.345(85)
^{168}Yb	0.13	1888.80(11)	3655.128(100)

good. Difficulties in frequency stabilisation arise when an isotope has a low natural abundance or if the atomic line significantly overlaps with the line of another isotope. Issues arising from lower abundance are fairly easily resolved by increasing the flux of the atomic beam by raising the temperature. The increase in oven temperature does result in a slight increase in the linewidth of the features but this is not significant. The other negative factor is the reduction in the lifetime of the Yb source.

The more challenging issue is the overlapping of lines for different isotopes. On the $^1S_0 \rightarrow ^1P_1$ transition this occurs for ^{172}Yb and ^{173}Yb $F' = 3/2$ and $F' = 7/2$ lines, in addition to ^{170}Yb and ^{171}Yb $F' = 1/2$ lines. Frequency stabilisation for loading a ^{173}Yb MOT on the $F = 5/2 \rightarrow F' = 7/2$ transition may be achieved by stabilising the laser to the ^{172}Yb peak and using the AOMs to account for the isotope shift. This works reasonably well because the ^{172}Yb line is much stronger than the neighbouring ^{173}Yb lines meaning the ^{172}Yb lock is relatively unperturbed. By looking only at the abundances presented in Table 3.3 it is not clear why the ^{172}Yb peak is much larger than the ^{173}Yb peaks, as both isotopes have similar abundances. However, ^{173}Yb possesses hyperfine structure so the transition strengths are also dependent on the Clebsch-Gordon coefficients which dilutes the intensity in comparison to bosonic lines.

The presence of hyperfine structure in the fermionic isotopes may be exploited to aid frequency stabilisation of the 399 nm laser to the ^{170}Yb transition. The relative strengths of the ^{170}Yb and ^{171}Yb $F' = 1/2$ lines are similar, which can cause the lockpoint to jump between the two peaks. The frequency stabilisation may be improved by rotating the polarisation of the 399 nm light to be vertical, parallel to the direction of fluorescence detection. The detected fluorescence spectra for two orthogonal cases of linear polarisation of the 399 nm laser for P = 265 μW are shown in Fig. 3.11a, where horizontal polarisation is shown in blue and vertical polarisation in red. The

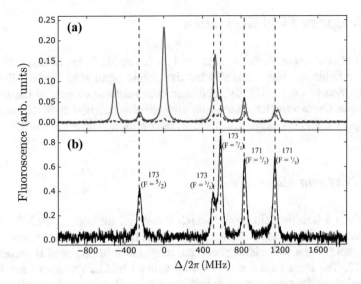

Fig. 3.11 a The fluorescence spectrum of the $^1P_1$ state for vertically (horizontally) polarised light is shown by the red (blue) line for a power $P = 265\,\mu\text{W}$ and temperature T = 470 °C. **b** Scaled difference of the two signals showing only the fermionic spectrum. The dashed vertical lines show the detunings measured by Das et al. [8]

vertical polarisation causes a suppression of the fluorescence for the bosonic isotopes due to their lack of ground state structure [42]. The fermionic spectrum obtained is used to stabilise the frequency to the ^{171}Yb $F' = 1/2$ peak more robustly. The isotopic frequency difference is again accounted for using AOMs. The pure fermionic spectrum is seen more clearly in Fig. 3.11b by subtracting the horizontally polarised signal from the vertically polarised signal (multiplied by a suitable factor of about 0.033 to match the heights of the bosonic peaks). We do not see a perfect extinction of the boson signal due to the large collection angle of the fluorescence detector.

Currently, using a few of the methods outlined above we are able to adequately lock to any transition required to load a MOT of almost any isotope (^{168}Yb is untested). The flexibility of the current locking scheme is useful for our current investigation of the many different mixtures of Cs and Yb, with the overarching goal to measure the scattering length. However, in the future, once the scattering lengths are known, we may opt to only use a specific Yb isotope. In this case a more robust scheme could be devised. This could involve stabilising the laser to the peak of ^{174}Yb because of its high SNR and then using additional AOMs to offset the frequency to match the isotope shift. This is feasible for the 399 nm transition but the large isotope shifts of the 556 nm transition would make this technique challenging. Another useful scheme, if fermionic ^{171}Yb is required, is to stabilise the 556 nm light to an inverted crossover resonance, which has been shown to produce a large dispersive locking signal for the MOT transition [30].

3.4   Magnetic Field Generation

Good control of magnetic fields is essential for the manipulation of ultracold atoms.
Magnetic fields are used when slowing the atomic beam effusing from the oven,
trapping atoms in the MOT, degenerate Raman sideband cooling of Cs, trapping
atoms in the Cs reservoir trap, tuning the intraspecies scattering length and absorption
imaging. The array of coils used to achieve this control are summarised in this section.

3.4.1   Zeeman Slower Coils

To produce a tunable field profile which is optimal for both Cs and Yb we use
a Zeeman slower wound with a set of five coils from three types of copper wire.
A schematic of the coil windings and the magnetic field produced is presented in
Fig. 3.12. The main field shape is set by Coils 1 and 2. The increasing field is
produced using the same current to both coils but with a reversed current direction
between the two, leading to two regions of opposite field. We use a slower with an
increasing field to ensure a large detuning of the Zeeman slower light in the MOT
region. This is critical when using the $^1S_0 \rightarrow{}^1 P_1$ Yb transition for Zeeman slowing
because the strength of the transition causes a significant pushing force on the Yb
MOT operated on the weak $^1S_0 \rightarrow{}^3 P_1$ transition. The Zeeman slower exhibits a
zero-crossing magnetic field profile which allows the use of smaller coil currents to
generate the $\simeq$540 G total field span. The large field span is required for Yb due to

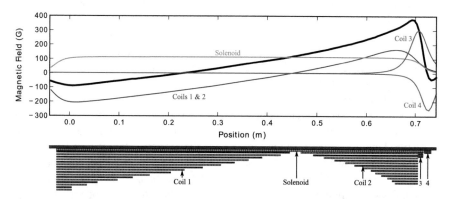

Fig. 3.12   A schematic detailing the Zeeman slower winding pattern. Turns from the 3 × 1 mm wire
are shown in green and red, the 4.3 mm wire turns are shown in yellow and the 3.4 mm wire turns
are shown in orange. The solenoid is wound directly onto a steel former tube (grey). The direction
of the atomic beam is from left to right. The turns in red (green) represent windings made from
the right (left) to the left (right) of the figure. Above the schematic is a plot showing the measured
Zeeman slower field profile when operating at Yb currents (black curve). The contribution to this
by each coil is also shown

the small wavelength of the Zeeman slower transition and the higher mean speed of the atoms effusing from the oven. Two small, high-current coils, Coils 3 and 4, are used to produce the large end field and the end field drop-off before the MOT region. Sufficient cancellation of the large end field is critical for Yb due to the small magnetic field gradient used. The solenoid coil allows the total field profile to be shifted up and down to match the Zeeman slower laser detuning required.

3.4.2 Science Chamber

As all experiments are performed on atoms trapped in the science chamber, the chamber is surrounded by an number of different coils to allow fine experimental control of the magnetic field. Figure 3.13a shows the current setup of the MOT and Bias coils within the re-entrant flanges of the science chamber.

Shim Coils

The shim coils are a versatile set of three pairs of coils used to apply small homogeneous fields up to 5 G inside the science chamber. The coils are oriented in three dimensions, East-West, North-South and Up-Down. The Up-Down shims are wound using 1 mm diameter wire directly onto the science chamber mount, whereas the other shim coils are wound using the same wire onto Tufnol formers and mounted on posts around the science chamber.

These coils are used to null stray magnetic fields in addition to the Earth's magnetic field. They are frequently used in the experiment for moving the quadrupole zero of the MOT to overlap with the dipole trap beams and for providing a small quantisation field during degenerate Raman sideband cooling or absorption imaging.

MOT Coils

The MOT coils are a pair of coils in the anti-Helmholtz configuration and are used to supply the quadrupole field for the MOT. The coils are formed of 16 turns of square cross-section tubing with a mean diameter of 75 mm and housed inside a nylon mount which is secured inside the upper and lower re-entrant flanges with a separation of 84 mm. The wire has a 2 mm bore running through the centre to allow the coils to be water-cooled. Aside from the MOT gradient, the coils are also used for magnetic levitation of Cs in the reservoir trap and levitated absorption imaging.

Bias Coils

The Bias coils are critical for the application of homogeneous magnetic fields which are larger than the 5 G provided by the shim coils. The current set of coils are formed from 4 turns of wire with a mean diameter of 86 mm. These coils are the 'fast Bias coils' described in the thesis of Kemp [20]. They are housed in the same nylon mount as the MOT coils and with the current electronic setup are capable of achieving fields of $\simeq$80 G. This field is much smaller than the expected fields at which interspecies CsYb Feshbach resonances of useful width are likely to occur [4, 5]. A set of coils

(a)   **(b)**

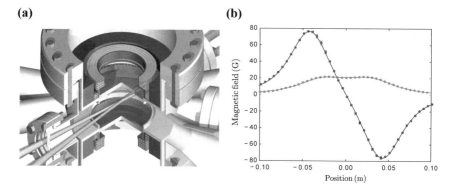

Fig. 3.13 a Rendering of the MOT and Bias coils within the re-entrant flanges of the science chamber. A representation of the dipole beams have been included as a reference to the centre of the chamber. **b** The measured axial magnetic field profile as a function of position of the Bias coil (blue markers) and quadrupole coils (red markers). The fields are produced with 30 A currents flowing through the coil pair in series. The solid lines show the fitted magnetic field profile using the Biot-Savart law

capable of reaching up to 2000 G have been designed but are yet to be installed in the setup.

Currently, the Bias coils are used for controlling the intraspecies scattering length of Cs during evaporation. A critical magnetic field strength which must be achievable is 22 G, the field at which three-body recombination loss is minimised [40], which is essential for the evaporation of Cs. The measured axial magnetic field profile of the MOT and Bias coils at a current of 30 A is shown in Fig. 3.13b. The measurement was performed outside the science chamber, with the coils clamped in their nylon mounts and set to the separation of the re-entrant flanges, 57 mm.

3.5 Absorption Imaging

3.5.1 Measuring the Density Distribution

The most common method to measure the density distribution of ultracold atoms is absorption imaging. In absorption imaging, the atoms are illuminated with a short pulse of resonant laser light and the shadow of the atoms is imaged onto a CCD array. Provided that the probe light does not saturate the transition, the fraction of light transmitted through the system decreases exponentially according to the Beer-Lambert law. The reduction in the initial intensity profile $I_0(x, y)$ of a beam propagating along the z–direction, through the atomic density distribution is described by

$$I(x, y) = I_0(x, y)\, e^{-n(x,y)\sigma_{\mathrm{tot}}}, \tag{3.1}$$

where $n(x, y)$ is the column density of the atomic cloud at position (x, y) and σ_{tot} is the absorption cross section of the transition. For a resonant, low light intensity ($I_0 \ll I_{sat}$) probe operating on a closed stretched transition of wavelength λ, the resonant absorption cross section is $\sigma_0 = 3\lambda^2/2\pi$ [10]. A CCD camera is used to detect the intensity profile of the initial and the attenuated probe beam. In the experiment, this is achieved by recording two sequential images, the first image with the light and the atoms present and the second image with no atoms, just the light. By comparison of these two images the column atomic density can be extracted

$$n(x, y) = -\frac{1}{\sigma_0} \ln\left(\frac{I(x, y)}{I_0(x, y)}\right) = \frac{1}{\sigma_0} OD(x, y). \tag{3.2}$$

In this equation OD (x, y) is the optical depth profile of the cloud.

When analysing thermal clouds in the experiment we use a fit function of the form

$$n(x, y) = n_0 \exp\left(-\frac{(x - x_0)^2}{2\sigma_x^2} - \frac{(y - y_0)^2}{2\sigma_y^2}\right), \tag{3.3}$$

where the fit parameters are the peak density n_0, cloud centres $x_0(y_0)$ and cloud widths σ_i.

From the measured density distributions it is possible to calculate the number of atoms in the cloud N, simply by integrating over the density distribution

$$N = \int\int n(x, y) \, dx dy = \frac{1}{\sigma_0} \int\int OD(x, y) \, dx dy. \tag{3.4}$$

A strong transition is favoured for absorption imaging as shorter probe pulses prevent 'smearing' of images due to atoms moving during the imaging pulse (like during a time of flight). A strong transition also allows a higher intensity of light to be used before saturation of the transition, this allows a higher SNR of counts on the CCD.

These requirements suggest that the $^1S_0 \rightarrow ^1P_1$ transition is the ideal choice for imaging of Yb atoms. Imaging of bosonic Yb is extremely simple as the lack of ground state structure means that all polarisations of light drive closed transitions. Therefore, when imaging bosonic Yb no quantisation magnetic field is required. This allows (near) simultaneous imaging of Cs and Yb atoms on the two separate cameras. The probe beam has a $1/e^2$ diameter of 13.6(2) mm and a typical power of $40\,\mu$W. This corresponds to an intensity of $I \simeq 1 \times 10^{-3}\,I_{sat}$, well in the low intensity regime. The magnification of the Yb imaging system is 1.16 resulting in a pixel size of 8.6 μm.

Absorption imaging of Cs is slightly more complex than Yb due to the presence of hyperfine structure. The only closed transitions are those of the stretched states, which requires good $\sigma^{+/-}$ polarisation of the imaging light. To accomplish this a quantisation field of 3.5 G is applied along the imaging axis of the probe beam.

The probe acts on the $F = 4 \rightarrow F' = 5$ transition and therefore needs repump light throughout the probe pulse to prevent the atoms from decaying into a dark state. The Cs probe beam has a $1/e^2$ diameter of 23.1(4) mm and we typically use a power of 200 μW which corresponds to an intensity of $I \simeq 0.1 I_{\text{sat}}$. The magnification of the Cs imaging system is 1.36 resulting in a pixel size of 11.8 μm.

Due to the high densities of the atoms after evaporative cooling it is often desirable to increase the time of flight (TOF) before taking an absorption image. The increase in expansion time causes the OD to fall and allows more reliable imaging of the atomic cloud. However, the TOF cannot be increased indefinitely as the total TOF is limited by the field of view of the imaging system. This effect may be counteracted by applying a levitating magnetic field gradient during the TOF which cancels the effect of gravity. To perform a levitated TOF, a bias field is used to shift the magnetic field zero below the atoms and at the same time a magnetic field gradient is provided by the MOT coils. A similar technique, called Stern-Gerlach separation, may also be used to separate the atoms into different magnetic sublevels. The different magnetic sublevels experience different forces due to their different magnetic moments. The gradient is applied for a period of time long enough to spatially separate the different sublevels and then a single absorption image is taken of the separated atom clouds. An example of a Stern-Gerlach experiment may be seen in Sect. 5.2.

3.5.2 Measuring the Momentum Distribution

In addition to the density distribution, absorption imaging is capable of measuring the momentum distribution of an atomic sample. This is achieved by allowing the atoms to freely expand for a TOF before the image is taken. A TOF measurement is performed by switching off all trapping potentials and allowing the atoms to fall under the influence of gravity. In the absence of all other potentials the atomic distribution evolves according to its initial momentum distribution.

For a thermal cloud of atoms expanding out of a harmonic trap, the temporal dependence of the cloud size is given by

$$\sigma_i(t) = \sqrt{\sigma_{i,0}^2 + \frac{k_B T}{m} t^2}. \tag{3.5}$$

By taking a TOF expansion series and fitting the size of the thermal cloud as a function of expansion time we may extract the temperature T and the initial size $\sigma_{i,0}$ of the cloud. However, in practice a measurement of the temperature may be performed in a single shot using

$$T_i = \frac{m}{k_B} \frac{\sigma_i^2}{1/\omega_i^2 + t^2} \tag{3.6}$$

where ω_i is the trapping frequency in the direction of the measured temperature.

References

1. Alcock C, Itkin V, Horrigan M (1984) Vapour pressure equations for the metallic elements: 298–2500 K. Can Metall Q 23(3):309–313. https://doi.org/10.1179/cmq.1984.23.3.309
2. Bjorklund GC (1980) Frequency-modulation spectroscopy: a new method for measuring weak absorptions and dispersions. Opt Lett 5(1):15–17. https://doi.org/10.1364/OL.5.000015
3. Boddy D (2014) First observations of Rydberg blockade in a frozen gas of divalent atoms. PhD thesis
4. Brue DA, Hutson JM (2012) Magnetically tunable Feshbach resonances in ultracold Li-Yb mixtures. Phys Rev Lett 108(4):043,201. https://doi.org/10.1103/PhysRevLett.108.043201
5. Brue DA, Hutson JM (2013) Prospects of forming ultracold molecules in $^2\Sigma$ states by magnetoassociation of alkali-metal atoms with Yb. Phys Rev A 87(5):052,709. https://doi.org/10.1103/physreva.87.052709
6. Butler KL (2014) A dual species MOT of Yb and Cs. PhD thesis
7. Cottrell T, Hultgren R (1973) Selected values of the thermodynamic properties of the elements. American Society for Metals, Metals Park, Ohio
8. Das D, Barthwal S, Banerjee A, Natarajan V (2005) Absolute frequency measurements in Yb with 0.08 ppb uncertainty: isotope shifts and hyperfine structure in the 399-nm $^1S_0 \to {}^1P_1$ line. Phys Rev A 72:032,506. https://doi.org/10.1103/PhysRevA.72.032506
9. De Laeter JR, Böhlke JK, De Bièvre P, Hidaka H, Peiser HS, Rosman KJR, Taylor PDP (2009) Atomic weights of the elements. Review 2000 (IUPAC technical report). Pure Appl Chem 75:683–800. https://doi.org/10.1351/pac200375060683
10. Foot CJ (2004) Atomic physics. Oxford University Press
11. Freytag R (2015) Simultaneous magneto-optical trapping of ytterbium and caesium. PhD thesis
12. Fukuhara T, Sugawa S, Takahashi Y (2007) Bose-Einstein condensation of an ytterbium isotope. Phys Rev A 76(5):051,604. https://doi.org/10.1103/PhysRevA.76.051604
13. Fukuhara T, Takasu Y, Kumakura M, Takahashi Y (2007) Degenerate Fermi gases of ytterbium. Phys Rev Lett 98(3):030,401. https://doi.org/10.1103/PhysRevLett.98.030401
14. Gehrtz M, Bjorklund GC, Whittaker EA (1985) Quantum-limited laser frequency-modulation spectroscopy. J Opt Soc Am B 2(9):1510–1526. https://doi.org/10.1364/JOSAB.2.001510
15. Guttridge A, Hopkins SA, Kemp SL, Boddy D, Freytag R, Jones MPA, Tarbutt MR, Hinds EA, Cornish SL (2016) Direct loading of a large Yb MOT on the $^1S_0 \to {}^3P_1$ transition. J Phys B At Mol Opt Phys 49(14):145,006. https://doi.org/10.1088/0953-4075/49/14/145006
16. Honda K, Takahashi Y, Kuwamoto T, Fujimoto M, Toyoda K, Ishikawa K, Yabuzaki T (1999) Magneto-optical trapping of Yb atoms and a limit on the branching ratio of the $^1P_1$ state. Phys Rev A 59(2):R934. https://doi.org/10.1103/PhysRevA.59.R934
17. Hopkins SA, Butler K, Guttridge A, Kemp S, Freytag R, Hinds EA, Tarbutt MR, Cornish SL (2016) A versatile dual-species Zeeman slower for caesium and ytterbium. Rev Sci Instrum 87(4):043109. https://doi.org/10.1063/1.4945795
18. Ishchenko V, Kochubei S, Rubtsova N, Khvorostov E, Yevseyev I (2002) Polarization echo spectroscopy of ytterbium vapor in a magnetic field. Laser Phys 12:1079–1088
19. Jayakumar A, Plotkin-Swing B, Jamison AO, Gupta S (2015) Dual-axis vapor cell for simultaneous laser frequency stabilization on disparate optical transitions. Rev Sci Instrum 86(7):073,115. https://doi.org/10.1063/1.4927198
20. Kemp SL (2017) Laser cooling and optical trapping of ytterbium. PhD thesis, Durham University
21. Kemp SL, Butler KL, Freytag R, Hopkins SA, Hinds EA, Tarbutt MR, Cornish SL (2016) Production and characterization of a dual species magneto-optical trap of cesium and ytterbium. Rev Sci Instrum 87(2):023105. https://doi.org/10.1063/1.4941719
22. Kim JI, Park CY, Yeom JY, Kim EB, Yoon TH (2003) Frequency-stabilized high-power violet laser diode with an ytterbium hollow-cathode lamp. Opt Lett 28(4):245–247. https://doi.org/10.1364/OL.28.000245

23. Kitagawa M, Enomoto K, Kasa K, Takahashi Y, Ciuryło R, Naidon P, Julienne PS (2008) Two-color photoassociation spectroscopy of ytterbium atoms and the precise determinations of s-wave scattering lengths. Phys Rev A 77(1):012,719. https://doi.org/10.1103/physreva.77.012719

24. Kleinert M, Gold Dahl ME, Bergeson S (2016) Measurement of the Yb I $^1S_0$–$^1P_1$ transition frequency at 399 nm using an optical frequency comb. Phys Rev A 94(052):511. https://doi.org/10.1103/PhysRevA.94.052511

25. Loftus T, Bochinski JR, Shivitz R, Mossberg TW (2000) Power-dependent loss from an ytterbium magneto-optic trap. Phys Rev A 61(051):401. https://doi.org/10.1103/PhysRevA.61.051401

26. Long Y, Xiong Z, Zhang X, Zhang M, Lü B, He L (2014) Frequency locking of a 399-nm laser referenced to fluorescence spectrum of an ytterbium atomic beam. Chin Opt Lett 12(2):021,401

27. Markowitz W, Hall RG, Essen L, Parry JVL (1958) Frequency of cesium in terms of ephemeris time. Phys Rev Lett 1:105–107. https://doi.org/10.1103/PhysRevLett.1.105

28. Maruyama R, Wynar RH, Romalis MV, Andalkar A, Swallows MD, Pearson CE, Fortson EN (2003) Investigation of sub-Doppler cooling in an ytterbium magneto-optical trap. Phys Rev A 68(011):403. https://doi.org/10.1103/PhysRevA.68.011403

29. McCarron DJ, King SA, Cornish SL (2008) Modulation transfer spectroscopy in atomic rubidium. Meas Sci Technol 19(10):105,601. https://doi.org/10.1088/0957-0233/19/10/105601

30. McFerran JJ (2016) An inverted crossover resonance aiding laser cooling of ^{171}Yb. J Opt Soc Am B 33(6):1278–1285. https://doi.org/10.1364/JOSAB.33.001278

31. Nemitz N (2008) Production and spectroscopy of ultracold YbRb* molecules. PhD thesis

32. Pandey K, Singh AK, Kumar PVK, Suryanarayana MV, Natarajan V (2009) Isotope shifts and hyperfine structure in the 555.8-nm $^1S_0 \rightarrow {}^3P_1$ line of Yb. Phys Rev A 80:022,518. https://doi.org/10.1103/PhysRevA.80.022518

33. Quessada A (2005) Development of an optical clock based on trapped strontium atoms: realization of an ultra-stable laser and frequency stability. PhD thesis

34. Shirley JH (1982) Modulation transfer processes in optical heterodyne saturation spectroscopy. Opt Lett 7(11):537–539. https://doi.org/10.1364/OL.7.000537

35. Steck DA (2010) Cesium D line data. http://steck.us/alkalidata (revision 2.1.4)

36. Sugawa S, Yamazaki R, Taie S, Takahashi Y (2011) Bose-Einstein condensate in gases of rare atomic species. Phys Rev A 84(1):011,610. https://doi.org/10.1103/PhysRevA.84.011610

37. Taie S, Takasu Y, Sugawa S, Yamazaki R, Tsujimoto T, Murakami R, Takahashi Y (2010) Realization of a SU (2) $\times$ SU (6) system of fermions in a cold atomic gas. Phys Rev Lett 105(19):190,401. https://doi.org/10.1103/PhysRevLett.105.190401

38. Takasu Y, Takahashi Y (2009) Quantum degenerate gases of ytterbium atoms. J Phys Soc Jpn 78(1):012,001. https://doi.org/10.1143/JPSJ.78.012001

39. Takasu Y, Maki K, Komori K, Takano T, Honda K, Kumakura M, Yabuzaki T, Takahashi Y (2003) Spin-singlet Bose-Einstein condensation of two-electron atoms. Phys Rev Lett 91(4):040,404. https://doi.org/10.1103/PhysRevLett.91.040404

40. Weber T, Herbig J, Mark M, Nägerl HC, Grimm R (2003) Bose-Einstein condensation of cesium. Science 299(5604):232–235. https://doi.org/10.1126/science.1079699

41. Xiong Z, Long Y, Xiao H, Zhang X, He L, Lv B (2011) A robust method for frequency stabilization of 556-nm laser operating at the intercombination transition of ytterbium. Chin Opt Lett 9(4):041,406

42. Zinkstok R, Van Duijn EJ, Witte S, Hogervorst W (2002) Hyperfine structure and isotope shift of transitions in Yb I using UV and deep-UV cw laser light and the angular distribution of fluorescence radiation. J Phys B At Mol Opt Phys 35(12):2693. https://doi.org/10.1088/0953-4075/35/12/305

Chapter 4
Quantum Degenerate Gases of Yb

The initial steps for the efficient production of ultracold molecules relies upon the preparation of an ultracold mixture of the constituent atoms, where both species are confined in a conservative trap. The numerous Yb isotopes that are capable of being cooled to quantum degeneracy [16, 17, 47, 48, 50, 51] make Yb an attractive element to work with in this regard. Quantum degeneracy in ultracold gases is normally achieved by evaporatively cooling the sample in a conservative trap. Bosonic Yb isotopes do not have a magnetic moment in the ground state and fermionic Yb isotopes only have a weak magnetic moment on the order of the nuclear magnetic moment, three orders of magnitude weaker than the Bohr magneton μ_B. Therefore, magnetic traps of fermionic Yb isotopes would require extremely large gradients >T/cm. A more sensible solution is the confinement of Yb in an optical dipole trap (ODT). This allows the magnetic field to be tuned without effecting the trapping potential, which enables the search for CsYb Feshbach resonances and magnetoassociation of the mixture.

In this chapter we describe the setup of an ODT for Yb and the development of cooling methods for the production of bosonic and fermionic degenerate gases of Yb.

4.1 Optical Trapping of Yb

Optical trapping is a pivotal tool in atomic and molecular physics that allows the confinement of magnetically-insensitive species (like Yb) in a trap suitable for evaporative cooling [19]. Optical dipole traps rely on the weak interaction of the atom's electronic dipole with the far-detuned light. During the interaction of an atom with laser light, the oscillating electric field of the laser $\mathbf{E}(\mathbf{r}, t)$ induces an atomic dipole

© Springer Nature Switzerland AG 2019   61
A. Guttridge, *Photoassociation of Ultracold CsYb Molecules
and Determination of Interspecies Scattering Lengths*, Springer Theses,
https://doi.org/10.1007/978-3-030-21201-8_4

moment $\mathbf{d}\,(\mathbf{r},t)$ that oscillates at the driving frequency ω. The amplitude of the dipole moment $d\,(\omega)$ is related to the amplitude of the electric field E_0 by

$$d = \alpha\,(\omega)\,E_0, \tag{4.1}$$

where $\alpha\,(\omega)$ is the complex polarisability of the ground state at frequency ω.

The interaction potential of the induced dipole moment $\mathbf{d}$ in the driving field E is given by

$$U_{\text{dipole}} = -\frac{1}{2}\mathbf{d}\cdot\mathbf{E}, \tag{4.2}$$

where the factor of $\frac{1}{2}$ is present due to the dipole moment being induced by the driving field as opposed to being a permanent atomic property. Taking the time average of Eq. 4.2 we obtain

$$U_{\text{dipole}}\,(\mathbf{r}) = -\frac{\text{Re}\,(\alpha)\,I\,(\mathbf{r})}{2\epsilon_0 c}, \tag{4.3}$$

where we have used $I = 2\epsilon_0 c |E_0|^2$ for the intensity of the light. When discussing the dipole trap it is common to refer to the peak value of the potential, U_0, as the trap depth.

It is important to note that the trap depth may be modified by additional terms such as the gravitational U_{grav} and magnetic potentials U_{mag}. In the vertical direction the total potential U_{tot} is given by

$$U_{\text{tot}} = U_{\text{dipole}} + U_{\text{mag}} + U_{\text{grav}}. \tag{4.4}$$

In the above equation we have incorporated the gravitational potential $U_{\text{grav}} = mgz$, where m is the mass of the trapped atom, g is the acceleration due to gravity and z is the vertical position. An illustration of the effect of gravity is shown in Fig. 4.1. The figure shows the variation of the optical trapping potential in the z direction. The red solid

Fig. 4.1 Simulation of an optical trapping potential for Yb atoms. The red solid line shows the combined optical trapping potential. The dashed blue (grey) line shows the dipole (gravitational) potential. The dotted red line shows the centre of the combined trapping potential

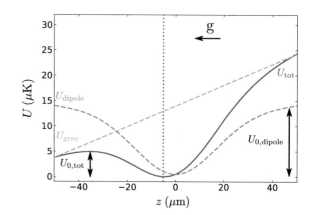

line is the combined gravitational (grey dashed) and dipole (blue dashed) potentials. The combined trapping potential is notably 'tilted' due to the gravitational potential, which has significantly reduced the depth of the trap in the vertical direction. In addition, the zero of the trapping potential is displaced vertically from the centre of trapping beam by the addition of the gravitational potential. This 'gravitational sag' of the trapping potential is given by $\Delta z = g/\omega_z^2$, where ω_z is the trapping frequency in the vertical direction. The zero of the trapping potential is shown by the red dotted line. The tilting of the trapping potential by gravity is very important for weak traps. It can be the basis for useful cooling techniques [23] or alternatively can be a problem which needs to be compensated. For example, differential gravitational sag between distinct atomic species can be troublesome when attempting sympathetic cooling [20].

In addition to the in-phase, dispersive component of the dipole oscillation that gives rise to the conservative dipole force on the atoms, it is also important to consider the out-of-phase, absorptive component that gives rise to the scattering force. The rate at which the atoms scatter the far-detuned light is given by

$$\Gamma_{\text{scat}}(\mathbf{r}) = \frac{1}{\hbar\epsilon_0 c}\text{Im}(\alpha)\,I(\mathbf{r}). \tag{4.5}$$

The trap depth and the scattering rate are two critical quantities in optical trapping. They are both a function of the position-dependent intensity $I(\mathbf{r})$ and the frequency-dependent complex polarisability $\alpha(\omega)$.

From Eq. 4.3 it is clear that the real part of the polarisability must be positive to form a potential well in which an atom can be trapped. In optical trapping a laser of frequency ω is used to form the trap, therefore we require knowledge of the sign of the frequency dependent polarisability $\text{Re}(\alpha(\omega))$ to inform our choice of trapping wavelength. In addition, the magnitude of the frequency dependent polarisability contributes to the design of the trap, as a large polarisability requires a smaller intensity of trapping light to produce a suitable trap. The frequency dependent polarisability of an atom in a specific atomic state may be calculated using the sum over states approach [38]

$$\alpha_0(\omega) = \frac{2}{3(2J+1)}\sum_{k\neq i}\frac{(E_k - E_i)\,|\langle\psi_i||D||\psi_k\rangle|^2}{(E_k - E_i)^2 - \omega^2}, \tag{4.6}$$

where $\langle\psi_i||D||\psi_k\rangle$ is the reduced dipole matrix element between states ψ_i and ψ_k. The weighted sum in Eq. 4.6 is a sum over all electronic transitions. However, in practice only the strongest transitions need to be considered at the level of precision we require. In more precise calculations such as the calculation of magic wavelengths [43], it is essential to include weaker transitions as the large number of weak transitions can contribute a few percent change in α [42].

Equation 4.6 gives only the scalar contribution to the polarisability from the valence electrons of the atom. In reality, there are also tensor contributions which

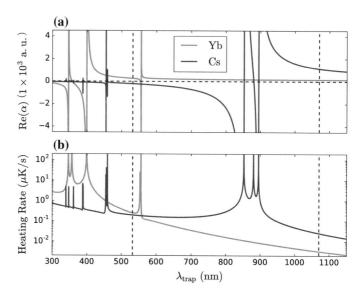

Fig. 4.2 Polarisabilities and heating rates for ground state Cs and Yb. **a** Real part of the frequency-dependent polarisability as a function of trap wavelength. **b** Heating rate as a function of trap wavelength for a trap depth of $U_0 = 50\,\mu K$. The vertical dashed lines show the commonly used trapping wavelengths of 532 and 1064 nm

are dependent on the orientation of the atomic angular momentum. We neglect these tensor contributions here as they are absent for ground state Cs and Yb atoms. As both Cs and Yb are widely used in atomic clock experiments, there exists a large amount of theoretical and experimental spectroscopic data for their ground states. We calculate α_0 using the transition frequencies and dipole matrix elements compiled in Ref. [43] for Cs and in the NIST database [1] for Yb.

Aside from the main contribution by the valence electrons of the atom, there are additional contributions by the ionic core and intermediate valence-excited states due to contributions from inner electrons [43]. In our calculations we include a static value for the core polarisability, using the value calculated in Ref. [24] and we neglect the very small core-valence contribution. The results of this calculation for the Cs $6S_{1/2}$ and Yb $(6s^2)^1 S_0$ states are shown in Fig. 4.2a.

It is also appropriate to consider the effect of the trapping light on the temperature of the atoms. In an optical dipole trap there is a heating contribution from the spontaneous scattering of trap photons. In the far-detuned case, each scattering event increases the thermal energy by twice the recoil energy. This arises because both the absorption and emission of a trap photon contribute $E_{rec} = \frac{k_B T_{rec}}{2}$, where $T_{rec} = \frac{\hbar^2 k^2}{m k_B}$. For a three dimensional trap the heating rate is [19]

$$\dot{T} = \frac{T_{rec} \Gamma_{scat}}{3}. \tag{4.7}$$

The heating rates for Cs and Yb are shown in Fig. 4.2b for a trap depth of $U_0 = 50\,\mu$K. The heating rate is strongest close to strong transitions for both species. For Cs this is in the 800–900 nm region near the D_1 and D_2 lines. For Yb, the dominant transitions are at much lower wavelengths, with a large heating rate in the 350–450 nm region close to the strong $^1S_0 \rightarrow {}^1P_1$ transition at 399 nm.

The wavelengths of commercially available high power lasers commonly used for optical dipole traps (532 nm and 1064 nm) are represented by vertical dotted lines in the figure. The heating rates are negligible for both species at both of these wavelengths due to the large detuning from resonance. At 532 nm the polarisability of Yb is positive but for Cs the polarisability is negative, meaning that the force is repulsive and atoms are pushed away from the focus of the laser. Fortunately, the polarisability is positive for both species at 1064 nm with values of $160\,a_0^3$ for Yb and $1140\,a_0^3$ for Cs. The Cs polarisability at this wavelength is seven times larger than Yb due to the smaller detuning from the strong infra-red Cs transitions compared to the strong Yb transitions which are in the UV region. A large difference in polarisability is not ideal for the simultaneous trapping of the two species but should suffice for the initial investigation of the interspecies scattering lengths. A more suitable trapping arrangement is possible by introducing an additional trapping beam at 532 nm. By tuning the intensity of the 532 nm beam along with the intensity of the 1064 nm beam we may create a balanced trapping potential for the two species.

4.1.1 Optical Setup

A 100 W fibre laser (IPG, YLR-100-LP-AC) that operates at a wavelength of 1070(3) nm is used to produce the light for the optical dipole trap. The high output power of the laser is required for optical trapping of Yb because of the Yb atom's low polarisability at this wavelength. Also, the absence of a magnetic moment in Yb necessitates loading of the dipole trap directly from the MOT without any initial evaporative cooling in a magnetic trap. Therefore, for optimal loading of the trap from the MOT, the trap depth produced must be an order of magnitude larger the temperature of the Yb MOT.

Figure 4.3 shows the setup of the optical dipole trap, which we shall label as the 'dimple' trap to distinguish it from the large volume dipole trap we employ later for Cs. A $\lambda/2$ waveplate and a PBS cube are placed just after the output of the IPG to allow the attenuation of the beam power during alignment. When operating at low laser powers there is a visible speckle pattern in the beam profile which is proportionally stronger at lower laser powers. The associated distortion of the beam profile makes measurement of the beam waist difficult at lower powers. It is likely this speckle pattern originates from light travelling down the cladding of the IPG fibre. The speckle pattern is unpolarised so the PBS also serves to improve the quality of the beam as it removes 50% of the speckle on the laser output.

After the PBS the beam size is reduced using a telescope composed of $f = 125$ mm and $f = -50$ mm lenses. The beam then passes through a high-power

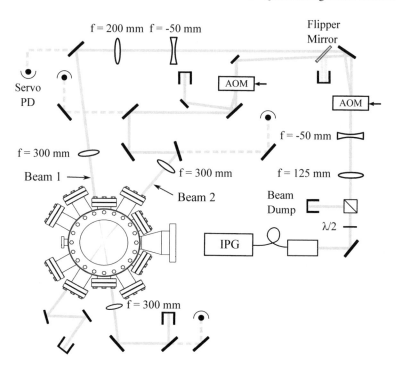

Fig. 4.3 Schematic showing the dimple trap optical layout

water-cooled AOM (Isomet, M1135-T80L-3) which is used for intensity stabilisation and experimental control of the trap depth. A flipper mirror is placed after the AOM and used for reflecting the 1st and 0th order light into a beam dump when required. With the flipper mirror retracted, the 1st order beam is then incident on another telescope used for beam shaping. Following the telescope the light passes through a $f = 300$ mm achromatic lens which focuses the light to a point inside the science chamber. We found that at locations where the beam radius was low, the use of high quality achromatic lenses instead of singlet lenses was essential to prevent significant thermal lensing. The $f = 300$ mm achromatic lens is mounted on a translation stage for precise adjustment of the waist position. The measured beam waist at the position of the atoms is 33(3) μm.

To maximise the power available for the trap and allow independent control of both beams, the 0th order from AOM 1 is used for the second arm of the crossed trap. The 0th order beam is picked off using a D shape mirror and subsequently passes through a second high-power water-cooled AOM. The 1st order of this second AOM is then incident on a $f = 300$ mm achromatic lens mounted on a translation stage. The measured waist of this second dimple beam is 72(4) μm and the crossing angle with beam 1 is 40°. We note that interference between the two dimple beams is not a concern as the large linewidth of the IPG laser corresponds to a coherence of length less than 1 mm.

A problem with the current setup is that we observe that the beam focus position significantly changes with time. This effect is due to thermal lensing of the optics and is prevalent in many high power systems like our own. To help mitigate some of the issues caused by thermal lensing, a flipper mirror was installed to allow the IPG laser and AOM 1 to be on during the Yb MOT load without the light entering the science chamber. Switching the laser on during the MOT load allows time for the lenses and the laser to thermalise, causing the focus to evolve to a constant position by the time we begin to load into the trap. Following the MOT load, the IPG laser is briefly switched off and the flipper mirror retracted before the laser is switched on again for dipole trap loading.

To evaporatively cool the trapped atoms, it is essential to precisely control the intensity of the trapping laser over a large range. We achieve this by stabilising the laser intensity using an AOM for each arm of the trap with independent PID feedback. A small fraction of the trapping light is transmitted through the back polished face of the mirrors used for alignment of the dimple beams and is detected by two servo photodiodes per trapping beam. The detected signals from the two photodiodes are summed and sent to an input of the PID control circuit. The summed voltage is compared to an external setpoint voltage from the LabVIEW control software and negative feedback is applied to the input of the AOM RF amplifier which controls the amount of RF power sent to the AOM. Light entering one of the photodiodes is attenuated by a large factor ($\sim$20) so the signals provided by the two photodiodes will differ by this attenuation factor. At high powers the unattenuated detector is saturated, contributing a constant offset to the PID and the attenuated detector is used to detect changes in the power. At low powers, the attenuated photodiode signal is very small but the unattenuated photodiode still has a moderately large voltage, which enables precise control of the laser intensity at low powers [15]. This two-photodiode system circumvents a problem typical for one-photodiode systems, where at low powers the detector signal is small and therefore vulnerable to electronic noise at mains frequencies, causing significant intensity noise on the trap laser.

4.1.2 Loading Yb into an Optical Trap

One slight complication of using Yb atoms in a cold atom experiment is that ground state ^{174}Yb atoms do not possess a magnetic moment. This requires the dimple trap to be loaded directly from a MOT as opposed to transfer from a magnetic trap like in most alkali atom experiments. For efficient loading into the dimple trap from a MOT, the MOT density must be maximised and the MOT must be cooled to low temperatures.

To load Yb into the dimple trap, we begin by loading an Yb MOT. The frequency of the MOT light is detuned by -4.5 MHz from the $^1S_0 \rightarrow \, ^3P_1$ transition and a total intensity of $I_{\text{total}} = 250 \, I_{\text{sat}}$ is used for the MOT beams. After 10 s of loading, we capture 5×10^8 atoms in the MOT at a temperature of 140 μK. Following loading of the MOT, the atomic beam shutter is closed to stop the flux of atoms and the Zeeman slower light and Zeeman slower magnetic fields are extinguished. The Zeeman slower light exerts a pushing force on the atoms in the MOT, even with a detuning of −578 MHz from resonance. Therefore, during the extinction of the Zeeman slower light and fields, the shim coils are changed to maintain the same MOT position in the absence of the Zeeman beam pushing force. The changes in the shim values also compensate for the change in the residual magnetic field at the atoms when the Zeeman slower coils are turned off.

Precooling Yb

Following the turn off of the Zeeman slower we initiate an intensity ramp of the MOT beams to cool the Yb MOT. The steady-state temperature of a two-level atom may be described by basic Doppler theory [14]

$$k_B T = \frac{\hbar \Gamma}{2} \frac{1 + 6I/I_{sat} + (2\Delta/\Gamma)^2}{4|\Delta|/\Gamma}, \qquad (4.8)$$

where k_B is Bolztmann's constant, T is the temperature, Δ is the detuning of the light and Γ is the natural linewidth of the transition. From Eq. 4.8 it is evident that a lower intensity of trapping light will reduce the temperature of the atoms in the MOT. However, there are subtle differences in the dynamics of a MOT operating on a weak optical transition in comparison to a broadband alkali MOT. In a broadband MOT the scattering force is much larger than the force exerted by gravity but in a MOT operated on a narrow transition, like the $^1S_0 \rightarrow {}^3P_1$ transition, the scattering force is comparable to the force due to gravity.

The differences between the two types of MOT are clear when the MOT beam intensity is reduced. The Yb MOT transition is power broadened during loading by many natural linewidths, making the linewidth of the transition comparable to an alkali MOT. However, by reducing the MOT beam intensity, the scattering force of the MOT light is reduced and the gravitational force begins to have a strong effect on the MOT shape. Gravity dominates at the centre of the trap due to the weakening of the scattering force and the atoms sag several millimetres under gravity until the force due to the upward propagating MOT beam balances the downward gravitational force. In the vertical direction the atoms interact only with the upward propagating MOT beam but in the horizontal direction they undergo free flight until they reach the boundaries of an ellipsoid. The axes of the ellipsoid correspond to positions where the magnetic field induced Zeeman shift balances the MOT beam detuning.

The sag of the Yb MOT at low MOT light intensities is shown in the series of absorption images and measurements presented in Fig. 4.4. Under low intensity conditions, the vertical MOT position is extremely sensitive to the MOT detuning [2, 35]. To produce the images in Fig. 4.4a we directly load a MOT of ^{174}Yb atoms using the 556 nm transition at an intensity of $I_{total} = 250\ I_{sat}$. After loading the MOT, we ramp down the intensity of the MOT light before taking an absorption image of the atoms using the strong $^1S_0 \rightarrow {}^1P_1$ transition.

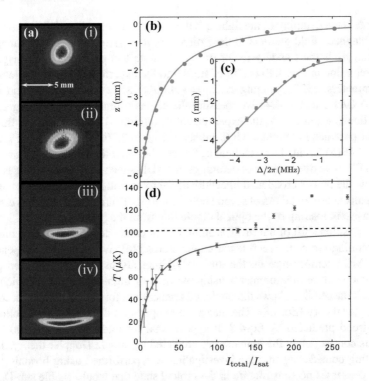

Fig. 4.4 Yb MOT behaviour as a function of MOT beam intensity. **a** Absorption images of the MOT for different total intensities of MOT light (i) 130 I_{sat} (ii) 38 I_{sat} (iii) 7.1 I_{sat} (iv) 1.7 I_{sat}. **b** Vertical sag due to gravity as a function of total intensity of the MOT light. The circles are experimental data and the green line is a fit of Eq. 4.9 with detuning $\Delta/2\pi = -4.6$ MHz and axial gradient $\frac{dB_z}{dz} = 3.4$ G/cm. **c** Vertical sag as a function of the detuning of the MOT light. The circles are experimental data and the green line is a fit of Eq. 4.9 with total intensity $I_{total} = 2.1\, I_{sat}$ and axial gradient $\frac{dB_z}{dz} = 3.4$ G/cm. **d** MOT temperature as a function of the total MOT beam intensity. The circles are experimental data, the black dashed line shows the temperature predicted by Doppler theory and the solid red line shows the temperature predicted by the modified Doppler theory

In Fig. 4.4b we measure the depth of the sag as a function of intensity and we see that the MOT drops a distance of up to 5 mm, which is greater than its initial size. The equilibrium z position may be determined by equating the force due to gravity with the net scattering force due to the two vertical MOT beams. This yields

$$
mg = \frac{\Gamma\hbar k}{2} \left[\frac{I/I_{sat}}{1 + 6I/I_{sat} + \frac{4}{\Gamma^2}\left(\Delta - \frac{\mu}{\hbar}\frac{dB_z}{dz}z\right)^2} \right.
$$
$$
\left. - \frac{I/I_{sat}}{1 + 6I/I_{sat} + \frac{4}{\Gamma^2}\left(\Delta + \frac{\mu}{\hbar}\frac{dB_z}{dz}z\right)^2} \right],
$$

(4.9)

where Δ is the detuning of the light, g is the standard acceleration due to gravity, $\frac{dB_{\hat{z}}}{dz}$ is the magnetic field gradient in the $\hat{z}$ direction, $\mu = g_J \mu_B$ is the effective magnetic moment, g_J is the Landé g-factor of the $^3P_1$ state and μ_B is the Bohr magneton. The green line in Fig. 4.4b is a fit of Eq. 4.9 to the experimental data (circles), with fit parameters: MOT detuning $\Delta/2\pi = -4.60(4)$ MHz and axial gradient $\frac{dB_{\hat{z}}}{dz} = 3.40(5)$ G/cm. In Fig. 4.4c we measure the sag as a function of detuning Δ. The green line is again a fit to the experimental data (circles) using Eq. 4.9, in this case with fit parameters of: total MOT intensity $I_{total} = 2.10(5) I_{sat}$ and axial gradient $\frac{dB_{\hat{z}}}{dz} = 3.40(5)$ G/cm. The excellent agreement between the experimental and fitted data in Fig. 4.4c over a range of detunings of 4 MHz shows that our detuning is stable to below the power-broadened linewidth of the transition and is highly reproducible during an experimental run of several hours duration. This reproducibility is essential for consistent loading of the optical dipole trap from the MOT.

The position shift of the MOT is an important parameter when utilising intensity and detuning ramps to reach lower temperatures [26]. We explore the dependence of the MOT temperature on the total MOT intensity in Fig. 4.4d. The red points show temperature measurements using the time of flight expansion technique and the black dashed line shows the predicted temperature for our parameters according to Doppler theory (Eq. 4.8). The measured temperature clearly does not follow the linear trend predicted by Eq. 4.8. It appears that for intensities below 100 I_{sat} the MOT is cooled below the temperature predicted by simple Doppler theory. This is surprising considering that the investigation was performed using bosonic ^{174}Yb, which possesses no substructure in the ground state that would enable sub-Doppler cooling. The origin of this additional cooling is linked to the shift in MOT position. As the MOT position drops for lower intensities of cooling light, the magnetic field experienced by the atoms changes due to the quadrupole gradient. The resulting Zeeman shift of the transition reduces the effective detuning experienced by the atoms. It is clear that as the MOT lowers, the atoms experience a detuning closer to resonance, which according to Eq. 4.8 results in a lower temperature. By finding the equilibrium z position of the atoms from Eq. 4.9 and using the effective detuning

$$\Delta_{eff} = \Delta - \frac{\mu}{\hbar} \frac{dB_z}{dz} z, \tag{4.10}$$

in Eq. 4.8 we obtain the modified Doppler theory curve (solid red line) in Fig. 4.4. This modified Doppler theory accurately predicts the temperature of the atoms for low MOT intensities but doesn't explain the observed temperatures at high MOT light intensities. This additional heating effect has been observed in Sr [56] and Yb [37] MOTs previously, with effects such as coherences between excited state levels [8] and transverse intensity fluctuations of MOT beams [6] proposed to be the cause.

The optimal intensity ramp for subsequent loading into the dimple trap is a ramp of the total intensity of the MOT beams from $I = 250 I_{sat}$ to $I = 4 I_{sat}$. At the end of the intensity ramp we typically obtain $N > 4 \times 10^8$ ^{174}Yb atoms at a temperature of 30 μK. A lower temperature is achievable using a lower number of atoms but this lower atom number reduces the amount loaded into the dimple trap. The increase in temperature with atom number is observed in other species [13, 25, 27] and is attributed to multiple scattering of photons within the atomic sample [3]. The final temperature achieved in the low density limit is also limited by the effective shaking of the trap due to the intensity noise on the MOT light [12].

To further increase the MOT density for dipole trap loading, we increase the MOT gradient from 2.5 to 7 G/cm and ramp the MOT detuning to a value close to resonance, from −4.5 to −1.5 MHz. The detuning ramp returns the MOT close to its original position before the intensity ramp and reforms into a more spherical shape. The increase in density and the better mode matching between the spherical cloud and the aspect ratio of the dimple trap beams enables more efficient loading into the trap. Detunings closer to resonance leads to a large loss of atoms due to the increase in light-assisted collisions [54] which are more pronounced for narrow optical transitions [12].

Dipole Trap Alignment

In the initial alignment of the dimple trap we used resonant 399 nm light copropagating with the 1070 nm trapping light. This allowed visible confirmation that the light was incident on the compressed Yb MOT. The resonant light exerted a strong pushing force on the MOT atoms causing the appearance of a hole in the MOT. Following the initial alignment of a single dimple trapping beam using the resonant light, we looked for evidence of atoms confined in the trap. To load the Yb MOT into the trap, we first compressed the Yb MOT using the intensity and detuning ramps described above and then switched on the trapping light for a period of a few 100 ms to allow loading into the trap. After this step the MOT light and gradient were extinguished to allow the untrapped atoms to fall away form the dimple trap region. Subsequently, the trapping light was switched off and an absorption image was taken after a brief 1 ms TOF. The appearance of a thin streak of atoms[1] signified confinement in the dimple trap. Further alignment of the trapping beam is performed by optimising the trapped number in the cloud by small horizontal and vertical adjustments to the final mirror before the chamber.

Alignment of the second beam is then performed by initially blocking the first (aligned) dimple trap beam and looking for evidence of trapping in the second beam; optimisation of the alignment is performed using the same technique as the first beam. The first beam is then unblocked and we observe a ball shape near the centre of the 'streak' corresponding to the crossing point of the two beams. The final number of atoms trapped is very sensitive to the trap depth, which in turn depends on the vertical overlap of the two beams, particularly at low overall trap depths. Hence, a final adjustment is performed by optimising the number of atoms remaining in the crossed region after a period of evaporation. The evaporation also aids alignment by removing the 'streak' of atoms confined in the single beam portion of the trap which can bias the results of the optimisation. More sensitive probes of the vertical

---

[1]Our imaging axis is near perpendicular to the propagation direction of the trapping light.

alignment are the performance of sympathetic cooling of Cs by Yb (discussed in Chap. 6) and the number loss during photoassociation of Cs and Yb discussed in Chap. 7).

For optimal optical trapping, the focus of both beams should overlap in the same region of the science chamber. However, alignment of the foci of the traps is difficult without the use of vertical imaging, which isn't present in our current setup. Instead, we use the two imaging axes we possess in the horizontal direction to check the overlap of the foci, the only complication being that one imaging axis can only image Cs and the other only Yb. This means optimisation of the alignment requires switching between loading Cs and Yb into the trap. For combined trapping of both these species the dimple trap beams are aligned to the Cs reservoir trap (more information in Sect. 5.3.1) using the above optimisation. This is because the Yb dimple trap loading can be optimised using the shims to move the MOT into the crossed region of the trapping beams, whereas the Cs cloud position is more difficult to translate as it occurs where four separate DRSC lattice beams are made to overlap. The dimple foci are then aligned through an iterative procedure of translating the beam 1 waist and realigning beam 1 until the centre of the Yb single beam trap is at the same position as the crossed beam region. The Rayleigh ranges of beam 1 and 2 are 3.2(6) mm and 15(2) mm respectively. The Yb atoms are used for this alignment because the imaging axis is closer to orthogonal with the beam 1 axis (see Fig. 4.3). The realignment of beam 1 is done by optimising the Cs load as an independent measure. The beam 2 focus can then be aligned by observing the Cs/Yb temperature after an evaporation ramp to a defined dimple beam power. The translation stage is set to the value corresponding to the highest trap depth which is the value that produces the highest temperature after evaporation. It is essential to check the alignment of beam 2 during this adjustment.

With the crossed trap optimally aligned we load 1.8×10^7 ^{174}Yb atoms into the dimple trap at a temperature of $T = 40 \, \mu$K for a power of 40 W and 15 W in beam 1 and 2 respectively. Initially, a large number of atoms are loaded into the single beam region of the trap, outside of the region where the two dimple trap beams overlap. The loading into the crossed region is enhanced by broadening the waist of beam 1 using AOM 1. We modulate the frequency of the AOM using a triangular wave with an amplitude of 3 MHz which corresponds to a time-averaged beam waist of $60(5) \, \mu$m at the atoms. The broadening enhances the loading by increasing the trap volume at the expense of the trap depth.

We observe that the number of atoms loaded into the dipole trap saturates as a function of time. Only for long trap loading times on the order of 5 s do we observe a decay in the number of atoms in the dimple trap. The saturation of the dimple trap load with time allows sufficient atoms to be loaded into the dimple even when the MOT number is low, just by holding the MOT over the dimple trap for longer.

Trap Frequency

Knowledge of the trapping frequency of the dimple trap is essential for devising evaporation routes to quantum degeneracy. Measurements of the trapping frequencies can be compared with our simulations of the potential in order to better understand

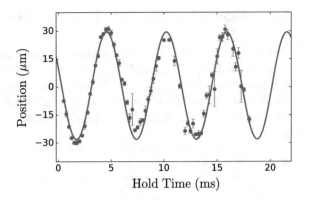

Fig. 4.5 Measurement of the trapping frequency for an ^{174}Yb BEC. The vertical displacement of the BEC after a 20 ms TOF is plotted as a function of time after a 'kick' is applied. The solid line is a sinusoidal fit to the data. The trap frequency extracted is 176(1) Hz for dimple beam powers 375 mW and 650 mW respectively

the trapping potential. The trap frequencies correspond to the motional frequencies of the atoms confined in a harmonic potential. There are multiple methods for measuring trapping frequencies, such as parametric heating, centre-of-mass (COM) oscillations and breathing mode oscillations. All are performed by dynamically changing the trapping potential to observe the effect on either the atom number, position or size of the trapped gas.

We typically use COM oscillations to measure the trapping frequencies. This involves displacing and releasing ('kicking') the trapped cloud of atoms and observing the oscillations of the cloud centre as a function of time. Figure 4.5 shows an example of measuring the vertical trapping frequency using an ^{174}Yb BEC. Measuring trapping frequencies in Yb is slightly more complex than in the alkalis as a magnetic field gradient cannot be used to initiate the oscillations. Instead, the atoms are set into motion by a sudden ramp up of the dimple trap power, this causes the atoms to oscillate back and forth in the trap. After a variable hold time the atoms are released from the trap and an absorption image is acquired. The trap frequency is extracted from a sinusoidal fit to the data such as the solid line shown in Fig. 4.5. For this dataset the trapping frequency extracted is 176(1) Hz and corresponds to a waist of 32.1(4) μm, in good agreement with our beam profiling measurements. Damping of the position oscillations may be observed if too large a kick is used to initiate the oscillations. If too large a kick is supplied, the atoms sample the anharmonic region of the trapping potential leading to a dephasing of the oscillations.

As the trap is approximately cigar shaped, the trap is characterised by radial and axial trapping frequencies. Measurements of COM oscillations in the axial direction are more difficult due to the angle of our imaging setup. Therefore, measurement of the axial trapping frequency is typically performed by the observation of width oscillations of the cloud, which are somewhat easier to observe.

4.2 Evaporative Cooling to Quantum Degeneracy

Bose-Einstein condensation is a new phase of matter which corresponds to macroscopic occupation of the ground state. This occurs when the temperature of the bosons falls below the critical temperature T_c. Below this temperature the de Broglie wavelength of the atoms is comparable to their spacing and the atoms form a coherent matter-wave. References [9, 28, 33] present extensive reviews on BECs in atomic gases.

The requirement for Bose-Einstein condensation of a gas is that the phase-space density (PSD) satisfies

$$\mathcal{D} = n_0 \lambda_{\text{dB}}^3 \geq 2.612, \tag{4.11}$$

where n_0 is the peak number density and $\lambda_{\text{dB}} = \sqrt{2\pi\hbar^2/mk_BT}$ is the de Broglie wavelength of the atoms. After loading of the dimple trap from the MOT we have a sample of 1.8×10^7 Yb atoms at a temperature of $40\,\mu\text{K}$. These atoms are distributed between the trapping region where both dimple beams cross and the region where trap is dominated by dimple beam 1. This makes estimation of a density difficult but roughly this corresponds to a PSD of 6×10^{-3}, far from the regime required for quantum degeneracy. Therefore, to reach the colder temperatures required to observe condensation we must further cool the atoms. The most efficient and widely used method for cooling atoms confined in an optical trap is forced evaporative cooling.

Evaporative cooling is extensively used in cold atom experiments to reach quantum degeneracy and an extensive review on this topic is available [29]. The evaporative cooling process involves cooling the atoms by truncating the Maxwell-Boltzmann distribution. To perform forced evaporation in an optical trap, the trap depth is lowered (usually by decreasing the intensity of the trapping light) causing the most energetic atoms to leave the trap. Elastic collisions between the remaining atoms allows the trapped atoms to thermalise and results in a thermal distribution with a lower mean temperature. Elastic collisions are critical for evaporative cooling. The elastic collision rate is given by

$$\Gamma_{\text{elastic}} = \langle n \rangle \sigma v = \frac{Nm\sigma\bar{\omega}^3}{2\pi^2 k_B T}, \tag{4.12}$$

where $\langle n \rangle$ is the mean density, v is the mean velocity, $\bar{\omega}$ is the mean trap frequency and σ is the scattering cross section. In the ultracold limit $\sigma = 8\pi a_s^2$ where a_s is the s-wave scattering length. As long as the rate of elastic (good) collisions is greater than the rate of inelastic (bad) collisions, the evaporative cooling process is efficient and the PSD of the sample increases as atoms evaporate. Therefore, a good figure of merit in evaporative cooling is the ratio of elastic to inelastic collisions or 'good to bad' collisions.

The efficiency of evaporative cooling γ is an important metric which measures the relative increase in the PSD and weights it against the number of atoms lost

$$\gamma = \frac{\ln \left(\mathrm{PSD}_f/\mathrm{PSD}_i\right)}{\ln \left(N_f/N_i\right)}. \tag{4.13}$$

The timing of the evaporation ramps are optimised for the greatest efficiency in each step. Efficient evaporation involves the careful management of the trapping potential and most importantly the trap frequencies. In magnetic traps runaway evaporation may be achieved with an increasing collision rate for each step [29], this is because the dominant loss mechanism in these traps is typically background gas collisions which is density independent. Therefore, increasing the collision rate reduces the time taken to thermalise and the ratio of good to bad collisions is increased. However, in optical traps where the density is much higher, thermalisation proceeds at a much faster rate. The higher densities in an optical trap introduce a new loss mechanism, three-body recombination [53], where two colliding atoms form a bound molecule and the third colliding atom carries away the binding energy, typically resulting on all three atoms being lost from the trap. Three-body recombination loss scales $\propto \langle n^2 \rangle \propto \bar{\omega}^6$. This means that an increasing collision rate will generally not lead to runaway evaporation as in a magnetic trap because three-body losses will eventually take over as the density increases.

Another important effect to consider in optical traps is the reduction in trapping frequencies throughout evaporation. Forced evaporation in optical traps is performed by reducing the intensity of the trapping light. This constant reduction of the trapping light intensity leads to a reduction in the trap frequencies, which results in a reduced collision rate and a longer rethermalisation time. Therefore, efficient evaporation in an optical trap is only achieved by careful management of the density throughout the evaporation.

4.2.1 Bose-Einstein Condensate of ^{174}Yb Atoms

Formation of an Yb BEC is a good test of our apparatus and experimental protocols. In order to reach the low temperature needed to form a BEC, we perform forced evaporation using a set of three linear ramps in the intensity of both dimple trap beams. These linear ramps have a decreasing gradient so as to approximate an exponential decrease in the trap depth. The full experimental sequence for the creation of an ^{174}Yb BEC is shown in Fig. 4.6, with the power of dimple beam 1 shown at the bottom of the plot.

During the evaporation of ^{174}Yb we start with a power of 40 W in beam 1 and 15 W in beam 2. We then ramp down the power of dimple beam 1 to 6 W and the power in beam 2 to 10 W over 3 s. The first ramp has the effect of increasing the number of atoms in the crossed region of the trap, as initially a significant fraction are confined in the single beam region. During this step we also ramp down the amplitude of the beam 1 AOM dither to zero, returning the beam 1 waist to 33 μm. We note that fast, efficient creation of BECs may be achieved using traps shaped dynamically

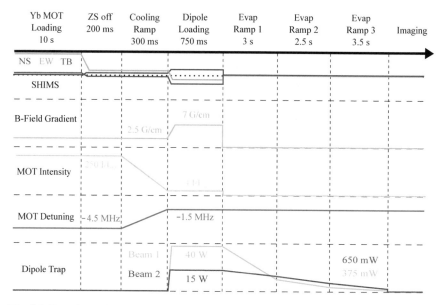

Fig. 4.6 Experimental sequence for the creation of an Yb BEC

throughout the evaporation [41], however, this introduces a large parameter space which is time consuming to explore experimentally. In Ref. [41] the authors found that an exponential power ramp and a linear ramp of the dither amplitude were among the optimum solutions, so in lieu of performing our own simulations we have utilised this scheme.

The majority of the evaporation is performed by reducing the beam 1 power, while the change in beam 2 power is significantly less. Due to the larger waist of the second beam it does not significantly contribute to the trap depth or trapping frequency in the radial dimension but it provides a significant enhancement in the axial trapping frequency. The purpose of the second beam is to maintain sufficient density near the end of the evaporation, where typically the collision rate drops due to the reduction in trap frequency with the lower trapping intensities. We note that a vertical ODT beam is used by Sr [46] and other Yb [11, 44, 51] experiments for the same purpose. The vertical beam offers greater independent control of the density as the beam does not significantly effect the trap depth in the vertical dimension where the evaporation occurs. Therefore, altering the power of the vertical beam only effects the trapping frequency axially along the horizontal beam and can be used to manage the density of the atoms in the trap, allowing a greater flexibility in the evaporation trajectory. However, the beam powers required in the single species experiments where it has been employed are typically much lower than those required for our 1070 nm trap, so due to safety concerns we opted for horizontal trapping beams.

Following the first evaporation ramp we ramp the powers in beam 1 from 6 to 1.5 W and in beam 2 from 10 to 4 W over 2.5 s. After this ramp we obtain 2.5×10^6

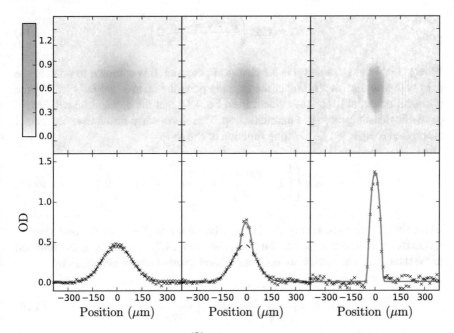

Fig. 4.7 Bose-Einstein condensation of ^{174}Yb by evaporative cooling. The top panels show absorption images taken after a 25 ms time of flight. The bottom panels show the corresponding horizontal cross-cuts through the images. The laser power of the optical dipole trap is gradually reduced, cooling the atoms from a thermal cloud at $T_{Yb} = 500$ nK (left), across the BEC transition to $T_{Yb} = 300$ nK (middle). Finally, at the end of evaporation a pure condensate (right) is produced containing 3×10^5 atoms

atoms at $T = 4\,\mu$K. Finally, we ramp the beam 1 power to 375 mW and the beam 2 power to 650 mW in 3.5 s.

Figure 4.7 shows the BEC transition in ^{174}Yb. The onset of quantum degeneracy occurs with $N = 9 \times 10^5$ atoms at a temperature of $T_c = 450$ nK. Further evaporation leads to the production of pure condensates containing 3 to 4×10^5 atoms. The measured trapping frequencies are 176(1) Hz radially and 29(1) Hz axially. The observed lifetime of the atoms is 6 s. This lifetime is shorter than the measured lifetime due to background gas collisions and may be due to some residual evaporation due to thermal lensing of the optics.

Imaging of a BEC

The properties of a BEC (or thermal cloud) are inferred from the density distributions measured using absorption imaging. It is essential when analysing the density distributions of a condensate that an appropriate model is used. A limiting case commonly applied to the analysis of condensates is the Thomas-Fermi limit. This limit allows the kinetic energy of the condensate to be neglected and is appropriate for $N a_s / a_{ho} \gg 1$, where $a_{ho} = \sqrt{\hbar/m\omega}$ is the harmonic oscillator length. In the Thomas-Fermi limit the density is the given by

$$n\left(\mathbf{r}\right) = \max\left(\frac{\mu - U_{\text{trap}}\left(\mathbf{r}\right)}{g}, 0\right). \tag{4.14}$$

Where $g = 4\pi\hbar^2 a_s/m$ and μ is the chemical potential. It is common to think of the density distribution as a filling of the trapping potential up to a 'height' given by the chemical potential [31]. It is evident from Eq. 4.14 that the condensate will have a parabolic density profile in a harmonic trap. Therefore, when fitting to experimental absorption images we use a fitting function of the form

$$n\left(x, y\right) = n_0 \max\left(\left[1 - \frac{(x - x_0)^2}{R_{\text{TF}, x}^2} - \frac{(y - y_0)^2}{R_{\text{TF}, y}^2}\right]^{3/2}, 0\right), \tag{4.15}$$

where the cloud centre $x_0(y_0)$, the Thomas-Fermi radius $R_{\text{TF}, i}$ and the peak density n_0 are free parameters in the fit. The green line in Fig. 4.7c shows a fit of this function to the data. The Thomas-Fermi radii are related to other useful quantities by

$$R_{\text{TF}, i}(0) = \sqrt{\frac{2\mu}{m\omega_i^2}} = a_{\text{ho}}\left(\frac{15 N_c a_s}{a_{\text{ho}}}\right)^{1/5}. \tag{4.16}$$

Therefore, from a fit of Eq. 4.15 to an absorption image we can extract the number of condensed atoms N_c and the chemical potential μ. To fit to bimodal distributions like the one shown in Fig. 4.7b we again use Eq. 4.15 but sum this with Eq. 3.3, the Gaussian distribution describing the thermal component of the gas.

Aspect Ratio Inversion in a BEC

The BEC transition is often easily identified by the striking increase in optical depth due to the macroscopic occupation of the zero-momentum state. Another tell-tale sign of condensation is the inversion of the aspect ratio after the condensate is released from a cigar shaped trap. Figure 4.8 shows the temporal evolution of the aspect ratio for a BEC and a thermal cloud. The evolution of the condensate half-widths $R_{\text{TF}, \rho}$ and $R_{\text{TF}, z}$ after release from a cigar shaped trap are given by

$$R_{\text{TF}, \rho}\left(t\right) = R_{\text{TF}, \rho}\left(0\right)\sqrt{1 + \tau^2}, \tag{4.17}$$

$$R_{\text{TF}, z}\left(t\right) = \left(\frac{\omega_\rho}{\omega_z}\right) R_{\text{TF}, \rho}\left(0\right)$$
$$\times \left(1 + \left(\frac{\omega_z}{\omega_\rho}\right)^2\left[\tau \arctan\tau - \ln\left(\sqrt{1 + \tau^2}\right)\right]\right), \tag{4.18}$$

where $\tau = \omega_\rho t$ and we have used ρ and z to label the radial and axial dimensions of the cigar shaped trap. The measurements of the aspect ratio ($R_{\text{TF}, \rho}/R_{\text{TF}, z}$) for an Yb BEC as a function of the time of flight are plotted in Fig. 4.8a alongside a fit using Eqs. 4.17 and 4.18. The large difference in aspect ratios between a BEC and

Fig. 4.8 Inversion of BEC aspect ratio during TOF. **a** Aspect ratio of BEC (green) as a function of TOF. The green solid line shows a fit of Eqs. 4.17 and 4.18 to the experimental data and the black dotted line shows the predicted behaviour of a thermal cloud. **b** Absorption image of a ^{174}Yb BEC after a TOF of 25 ms. **c** Absorption image of thermal ^{174}Yb atoms after a TOF of 25 ms

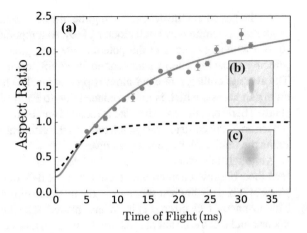

a thermal cloud are seen in the absorption images Fig. 4.8b, c. The thermal cloud expands isotropically for $t > \omega_z^{-1}$ as the mean velocity is isotropic in a thermal gas. However, for the BEC the expansion is faster in the axis of tight confinement due to the anisotropic release of the mean-field energy which is dictated by the shape of the trapping potential.

4.3   Degenerate Fermi Gas of ^{173}Yb Atoms

So far we have only considered the bosonic ^{174}Yb isotope in our discussion of trapping and cooling of Yb to quantum degeneracy. However, Yb possesses numerous stable isotopes which can be trapped and cooled, including two fermionic isotopes, ^{173}Yb and ^{171}Yb. Due to its larger scattering length ($a_{173} = 199\,a_0$ versus $a_{171} = -3\,a_0$) [32], ^{173}Yb is the more suitable fermionic isotope for the creation of a degenerate Fermi gas (DFG) and is used in the majority of fermionic Yb experiments [4, 11, 17, 45].

Evaporatively cooling fermionic atoms to degeneracy is complicated by the Pauli principle which states that identical fermions do not undergo s-wave collisions. As the gas is evaporatively cooled below the p-wave barrier, the elastic collision rate (required for evaporation) becomes extremely small. Efficient evaporative cooling of fermions may be achieved using sympathetic cooling with another species or isotope. The ^{171}Yb isotope possesses an extremely small scattering length that requires it to be cooled sympathetically. A degenerate Fermi gas of ^{171}Yb has been achieved using another Yb isotope [17] or Rb [52] as a coolant. The other alternative for evaporative cooling of a fermion is using a mixture of different internal states. ^{173}Yb has an s-wave scattering length of $a = 199\,a_0$ and a large nuclear spin of $I = 5/2$ which is well suited for evaporative cooling in a mixture of spin states.

As the gas enters the degenerate regime, the final temperature achievable through evaporation is limited by Pauli blocking [10]. At temperatures $T < 0.5\,T_F$, the majority of low-energy states in the potential are occupied. Due to the Pauli exclusion principle these low energy states cannot be doubly occupied as a result of a collision. The average collision rate per atom is proportional to the density of final states the atom can access, which is much reduced due to Pauli blocking of the lower energy states. Therefore, the collision rate is reduced and tends toward zero as $(T/T_F)^2$. The reduction in the collision rate much reduces the efficiency of evaporative cooling and limits the final achievable temperature.

Although the evaporative cooling of fermionic Yb atoms is more challenging than their bosonic counterparts, there is a great deal of interest in degenerate Fermi gases of Yb. This interest stems from the nuclear spin of the fermionic Yb isotopes. The absence of electronic spin in the ground state leads to a decoupling of the nuclear and electronic angular momenta, making atomic interactions independent of nuclear spin state. Therefore, the six nuclear spin states of ^{173}Yb exhibit an effective SU($N = 2I + 1 = 6$) spin-symmetry of the interaction Hamiltonian. The enlarged symmetry present in this system can lead to exotic correlated ground states and topological excitations [5, 7, 18, 21, 22, 55]. In addition, the nuclear spin can act as an entropy bath for the motional degrees of freedom in a Pomeranchuk cooling type effect [49]. This effect aids the observation of antiferromagnetic spin correlations of fermions trapped in an optical lattice [40]. Also, the existence of the long-lived clock state in Yb allows the manipulation of spin-orbit interactions in these systems using a synthetic (electronic) dimension [34, 36, 45].

The main reason behind our own interest in ^{173}Yb is the strong promise of observing interspecies Feshbach resonances in fermionic Yb isotopes compared to the bosonic isotopes. For subsequent investigations of Feshbach resonances, a high phase-space-density sample of ^{173}Yb is desirable.

4.3.1 Fermionic MOT

All bosonic isotopes of Yb have nuclear spin $I = 0$ and therefore only have a single magnetic substate in the $^1S_0$ ground state. However, for fermionic Yb isotopes there are many magnetic substates in the ground state due to the presence of nuclear spin, $I = 1/2$ for ^{171}Yb and $I = 5/2$ for ^{173}Yb. The magnetic moment of these fermionic isotopes is given by the nuclear magnetic moment μ_N, which is weaker than the Bohr magneton μ_B by a factor of $\sim$2000. This leads to a very small Zeeman splitting of the ground state magnetic sublevels, with the Zeeman shift of the MOT transition dictated by the magnetic moment of the $^3P_1$ excited state. This leads to differing forces depending on the m_F state of the atom. MOT light of a given frequency may be resonant, blue-detuned or red-detuned with respect to the atoms, corresponding to forces pushing some atoms toward the trap centre but also a force pushing some atoms away.

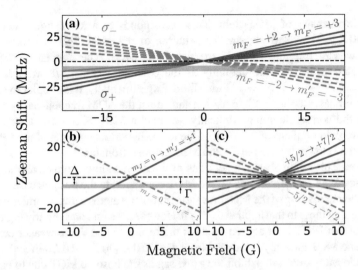

Fig. 4.9 Effect of magnetic substructure on MOT operation. **a** Zeeman shifts of the MOT transition for an alkali MOT. The ^{87}Rb $F = 2 \rightarrow F' = 3$ transition is used for illustrative purposes. The five allowed σ^+ transitions and the five allowed σ^- transitions are shown by the red solid and blue dashed lines respectively. The grey bar shows the frequency of the MOT light. **b** MOT diagram for bosonic Yb. **c** MOT diagram for fermionic ^{173}Yb ($I = 5/2$). A given frequency of MOT light may be red- or blue-detuned for different Δm_F transitions

The contrast between a fermionic MOT and a standard alkali MOT is shown in Fig. 4.9. Figure 4.9a shows the Zeeman shift of the MOT transition for an alkali MOT. In the example shown, all σ^- transitions have a negative Zeeman shift in a positive magnetic field and all σ^+ transitions have a positive Zeeman shift for positive magnetic field. Therefore, all m_F states experience a restoring force toward the centre of the trap and the value of magnetic field which Zeeman shifts the transition into resonance with the MOT light is relatively independent of m_F state. This behaviour remains for the very simple case of the bosonic Yb MOT shown in Fig. 4.9b. However, for the fermionic ^{173}Yb MOT shown in Fig. 4.9c, the magnetic field required to shift the transition into resonance is very sensitive to the m_F state of the atom. As a consequence, some m_F states are resonant with the anti-restoring MOT beam, causing a pushing out of the trap. This can be seen in the diagram where red (blue) lines have a negative Zeeman shift for positive (negative) magnetic fields.

Fortunately, steady-state operation of a MOT is possible due to the favourable transition probability difference between σ^+ and σ^- transitions for a given m_F state. The optical pumping in Yb is sufficiently fast that a sufficient average restoring force is produced without the use of a stirring laser to randomise the m_F states [39]. Due to the averaging of the restoring force over many m_F states, the trapping force is reduced in fermionic MOTs compared to bosonic MOTs.

To load a MOT of fermionic ^{173}Yb atoms the MOT light is detuned by -3.2 MHz from the $^1S_0$, $F = 5/2 \rightarrow {}^3P_1$, $F' = 7/2$ transition and we use the maximum available intensity of MOT light. This corresponds to a MOT light intensity of $I = 250\,I_{\text{sat}}$. The relatively smaller detuning of the MOT light is necessary due to the weaker trapping force in the fermionic MOT. The Zeeman slower light is operated with the same detuning as the bosonic MOT, -578 MHz from the $^1S_0, F = 5/2 \rightarrow\,^1P_1, F' = 7/2$ transition. Experimentally, it is simpler to stabilise the 399 nm laser to the ^{172}Yb transition and adjust the AOM frequencies by 58 MHz to give the correct detuning. Typically, we load a MOT of 5×10^7 atoms with a magnetic field gradient of 3.1 G/cm. Due to the hyperfine splitting of the $^1P_1$ state, the Zeeman slower light is only -262 MHz detuned from the $F = 5/2 \rightarrow F' = 5/2$ transition. This relatively small detuning of the Zeeman light from an atomic transition leads to an increased force on the MOT compared to the bosonic case.

Following loading of the MOT, we further cool the atoms using a similar intensity and detuning ramp to that applied to ^{174}Yb. However, we find that slightly higher final intensities of MOT light are required for the fermions due to the weaker confining force. The MOT intensity is reduced from $I = 250\,I_{\text{sat}}$ to $I = 10\,I_{\text{sat}}$ in 300 ms. The fermionic MOT does not exhibit the same sag as the bosonic MOT due to each m_F state becoming resonant with the MOT light at different positions from the MOT centre. At the end of the ramp the atoms reach a temperature of 25 μk which is lower than in the bosonic isotopes due to presence of sub-Doppler cooling in Yb isotopes with nuclear spin [37]. Following the intensity ramp, the atoms are compressed with a ramp of the magnetic field gradient to 6 G/cm and subsequently overlapped with the dipole trap position using the shims to displace the quadrupole zero. We load 6×10^6 ^{173}Yb atoms into the optical dipole trap, a good starting point for evaporative cooling.

4.3.2 Fermi Degeneracy in ^{173}Yb

Thermometry of an Ideal Fermi Gas

To measure the temperature of a weakly-interacting Fermi gas, we release the ^{173}Yb atoms from the trap and absorption image the atoms after a time of flight. It is convenient to use a fitted radius of the cloud R_i [30] where

$$R_i^2 = \frac{2k_B T}{m\omega_i^2} f(\xi),\tag{4.19}$$

$$f(x) = \frac{1 + x}{x} \ln(1 + x).$$

The value of R_i is directly related to the physical size of the cloud and allows us to interpolate between the classical regime ($T/T_F \gg 1$) where the cloud size is related to the thermal distribution of the atoms and the degenerate regime where the cloud size saturates at the Fermi radius. The fugacity is defined as $\xi = e^{\mu/k_B T}$ where μ is the chemical potential.

The absorption image is analysed using a 2D fitting function [30]

$$n(x, y) = n_0 \frac{\text{Li}_2\left(\pm \exp\left[\ln \xi - \left(\frac{(x-x_0)^2}{R_x^2} + \frac{(y-y_0)^2}{R_y^2}\right) f(\xi)\right]\right)}{\text{Li}_2(\pm \xi)}, \tag{4.20}$$

where $\text{Li}_m(x) = \sum_{k=1}^{\infty} \frac{x^k}{k^m}$ is a polylogarithmic function of order m and n_0, ξ, x_0, y_0 and R_i are fit parameters. Initial guesses for x_0, y_0 and R_i are made using a 2D Gaussian fit to the data. This aids the fitting as the calculation of the polylogarithmic function is fairly time consuming during 2D fits.

The degeneracy of the sample is characterised by the fugacity. For small fugacities the fitting function reduces to a Gaussian distribution of a thermal cloud and for high fugacity the fit function tends to the zero-temperature Fermi-Dirac distribution. The T/T_F parameter can be derived from the fugacity

$$\frac{T}{T_F} = [-6\text{Li}_3(-\xi)]^{-1/3}. \tag{4.21}$$

Therefore, we may extract T/T_F directly from the fit to the absorption image. The subtle change in the density distribution that is the hallmark of Fermi degeneracy is in stark contrast to the BEC case, where degeneracy is signalled by a dramatic enhancement of the OD. The difficulty in detecting Fermi degeneracy is highlighted in Fig. 4.10. The figure shows a comparison between Gaussian and Fermi-Dirac distributions at increasing levels of degeneracy. The difference between the two distributions is not perceptible at $T/T_F = 0.8$ and is only of the order of a few % at $T/T_F = 0.2$. It is easy to see that the T/T_F value extracted from the fit may be artificially enlarged due to non-ideal aspects of the imaging. Effects such as finite resolution, saturation and out of focus imaging may wash out the non-Gaussian features of the distribution, increasing the extracted T/T_F.

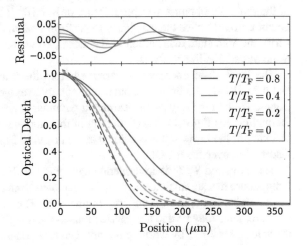

Fig. 4.10 Theoretical radial profiles of a degenerate Fermi gas. Optical depth is plotted as a function of radial distance from the centre of the cloud for various values of T/T_F. The dotted lines show the result of fitting a Gaussian to the Fermi-Dirac distributions (solid lines). The residuals from the Gaussian fit are shown in the upper section of the plot

The degeneracy of the Fermi gas may also be found from knowledge of the trap frequencies, the number of atoms and their temperature. The Fermi temperature is given by

$$T_F = \frac{\hbar\bar\omega\,(6N_i)^{1/3}}{k_B} \tag{4.22}$$

where N_i is the number of atoms per spin state. The temperature may be obtained from a fit to the wings of the density distribution, beyond the Fermi radius. For very low temperature clouds thermometry is difficult due to the low SNR of the wings. The SNR in the wings can be enhanced by azimuthally averaging over the 2D distribution. Integration over a single radial direction should be avoided as this will mix regions of different local T/T_F.

Evaporative Cooling of ^{173}Yb to Fermi Degeneracy

Starting from our sample of 6×10^6 ^{173}Yb atoms trapped in the dipole trap, we ramp down the power in dimple beam 1 from 40 to 6 W over 2.5 s. Over this same time period, the power in dimple beam 2 remains at 10 W. The second evaporation ramp involves ramping the power in beam 1 to 1.5 W and beam 2 to 4 W over 1.5 s. During this ramp the dither applied to the beam 1 AOM is reduced to zero and the waist of dimple beam 1 reduces from 60(5) μm to its original size 33(3) μm. The timescale for the first two evaporation ramps here is slightly shorter than the ramps used for ^{174}Yb due to the higher scattering length of ^{173}Yb reducing the necessary rethermalisation time during evaporation.

Following ramp 2 the gas is at $T = 700$ nK with 3×10^5 atoms in the sample. In the third evaporation ramp the dimple beam 1 and 2 powers are lowered to 420 mW and 710 mW respectively over 3 s. At the end of this ramp we have 2×10^5 atoms at $T = 350$ nK. The gas is cooled into the degenerate regime with a further power ramp by reducing the beam 1 power to 320 mW and the beam 2 power to 620 mW. The time for rethermalisation is extended in the final ramp which takes 4.5 s, this is due to a decrease in elastic collision rate by Pauli blocking of collisions. At the end of the ramp we obtain a six-spin mixture of 8×10^4 ^{173}Yb atoms at $0.4\,T/T_F$. We do not currently have the capability to perform optical Stern-Gerlach separation to verify the spin state distribution of the atoms, so we have assumed that the sample is unpolarised in this analysis.

The level of degeneracy is determined from the fitted fugacity using Eqs. 4.15 and 4.21. To aid the fitting we have azimuthally averaged over the 2D density distributions, the resulting fit and extracted parameters are shown for degenerate and thermal clouds Fig. 4.11. The flattening of the momentum distribution due to Pauli blocking of the lower states is evident at the origin of Fig. 4.11a where the Gaussian distribution over-fits the OD.

The extracted T/T_F is in moderate agreement with the value of $0.3\,T/T_F$ from temperature measurements, where T_F is calculated from Eq. 4.22 using the measured atom number per spin state and trap frequencies. The discrepancy between the two measurements may be due to imperfections in the imaging system which can inflate the value over T/T_F by washing out non-Gaussian features.

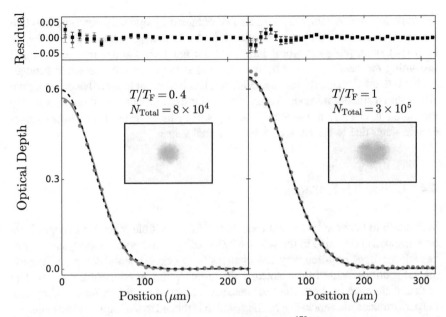

Fig. 4.11 Creation of a six-component degenerate Fermi gas of ^{173}Yb. The plot shows azimuthal averages of the optical depth as a function of radial distance from the cloud centre. The solid (dashed) line shows the azimuthally averaged 2D fit of a Fermi-Dirac distribution (Gaussian distribution) to the images. The upper plot shows the residuals from the Gaussian (black) and Fermi-Dirac fits (green). The insets show 15 ms TOF absorption images of degenerate (left) and thermal (right) atoms

Over the course of our work with ^{173}Yb, it became apparent that our current imaging setup is not ideal for reliable thermometry of Fermi gases. The imaging system was setup for use with bosonic isotopes, which, due to the lack of substructure in the ground state, have perfectly closed transitions for all polarisations of imaging light. However, the presence of ground state structure in fermionic Yb requires a well circularly polarised probe beam. The circularly polarised probe pumps the atoms, initially in an ensemble of spin states, into a stretched m_F level which allows imaging on a closed $\sigma^{+/-}$ transition like in the imaging of alkali atoms. At the time the data was taken we did not have the appropriate polarisation optics for circularly polarising the imaging light and the green MOT light along the imaging axis. Poor polarisation of the imaging light leads to a reduction in the OD of the absorption images, which leads to poorer fits and an undercounting of the atom number. In addition, the 8.6 μm resolution of imaging system is not sufficient to resolve the % level discrepancy between the Gaussian and Fermi-Dirac fits for small atom clouds where longer times of flight are not effective in improving the fits because of the reduction in optical depth. It is therefore very likely that the T/T_F is overestimated in the presented fits and improvements in the imaging setup would lead to better thermometry of the atoms.

More degenerate samples could also be obtained by optically pumping into two (or more) spin states following evaporation. As T_F is dependent on the number in each spin state, pumping atoms into a smaller number of states increases T_F and (assuming the heating caused by the pumping is not too large) the gas is brought deeper into the degenerate regime. However, it is important to note that pumping the atoms into a smaller number of spin states prior to evaporation will not necessarily reduce the final T/T_F as the Pauli blocking will be more severe for a lower number of spin states due to the reduced density of final states.

4.4 Other Yb Isotopes

In addition to bosonic ^{174}Yb and fermionic ^{173}Yb, multiple other Yb isotopes have been trapped and cooled in the setup. MOT loading and dipole trapping using other bosonic isotopes requires very few changes to the experimental setup and experimental routines, aside from relocking the lasers to the appropriate transition. The critical difference between the final densities achieved for the different isotopes is the performance of evaporative cooling, which is related to the intraspecies scattering length.

To prepare the second most abundant Yb isotope, ^{172}Yb, the MOT and dipole loading is performed using the same routine as ^{174}Yb except for a slightly longer MOT load time. The same number of atoms are prepared in the dipole trap as ^{174}Yb and the initial evaporation occurs very quickly due to the large scattering length $a_{172} = -598\,a_0$. However, at temperatures $T < 4\,\mu K$ there is a large loss of atoms due to the large three-body recombination rate (which scales $\propto a^4$). The strong attractive interactions makes formation of a stable ^{172}Yb condensate impossible for $N > 100$. For this reason, ^{172}Yb is the only Yb isotope that has not been cooled to quantum degeneracy.

The next most abundant bosonic isotope is ^{176}Yb which has a scattering length of $a_{176} = -24\,a_0$. Again, the loading into the dipole trap is achieved using the same procedure as ^{174}Yb. The lower abundance requires a longer MOT load time and an increased dipole trap loading time but the result is a near identical load of atoms into the dipole trap. Evaporative cooling is slow due to the much smaller scattering length and samples of 3×10^5 atoms at $4\,\mu k$ have been produced.

The properties of ^{170}Yb are relatively favourable for evaporative cooling. This isotope has a factor of 10 smaller abundance than ^{174}Yb but a moderate scattering length of $a_{170} = 64\,a_0$ that is favourable for evaporative cooling. Unfortunately, both $^1S_0 \to {}^1P_1$ and $^1S_0 \to {}^3P_1$ transitions of this isotope overlap very closely with another Yb isotope which complicates stabilisation of the cooling lasers. For the narrower $^1S_0 \to {}^3P_1$ transition this is not so troublesome, however, for the broader $^1S_0 \to {}^1P_1$ transition the splitting between isotopic lines is commensurate with the transition linewidth. To stabilise ZS light for a ^{170}Yb MOT we eliminate the fluorescence signal from the bosonic isotopes by rotating the polarisation of the spectroscopy light (as discussed in Sect. 3.3.2). The 399 nm laser is then stabilised to the remaining ^{171}Yb $F = 1/2$ peak. To maintain the same Zeeman slower detuning used for the other bosonic isotopes the ZS AOM frequencies are offset by the isotope shift, 40 MHz.

The number of ^{170}Yb atoms loaded into the dipole trap is slightly lower than the previously mentioned bosonic isotopes due to its lower abundance. However, evaporative cooling is fairly efficient and by extending the length of the ^{174}Yb evaporation ramps appropriately we find that we can produce ^{170}Yb BECs containing 1×10^5 atoms.

The remaining bosonic isotope, ^{168}Yb, has a very low natural abundance (0.13%). We have not seriously pursued creation of a ^{168}Yb MOT due to the difficulty in stabilisation of our lasers to the weak atomic spectroscopy signal and the long loading times that would be required. However, BECs of this isotope have been produced in Kyoto [47] and should be achievable if required.

Cooling of fermionic ^{171}Yb to degeneracy requires a sympathetic coolant due to its extremely low scattering length $a_{171} = -3\,a_0$. We have loaded atoms into the dipole trap from a ^{171}Yb MOT but the loading isn't as efficient as with the bosonic isotopes. Further experimental investigation is required to improve the loading efficiency of the dipole trap from fermionic MOTs. We defer investigation of this isotope until measurements of the Cs+Yb scattering length elucidate the viability of sympathetic cooling with Cs.

References

1. Kramida A, Ralchenko Yu, Reader J, NIST ASD Team (2018) NIST Atomic Spectra Database (ver. 5.6.1). National Institute of Standards and Technology, Gaithersburg, MD. https://physics.nist.gov/asd
2. Boddy D (2014) First observations of Rydberg blockade in a frozen gas of divalent atoms. PhD thesis
3. Boiron D, Michaud A, Lemonde P, Castin Y, Salomon C, Weyers S, Szymaniec K, Cognet L, Clairon A (1996) Laser cooling of cesium atoms in gray optical molasses down to 1.1 μk. Phys Rev A 53:R3734–R3737. https://doi.org/10.1103/PhysRevA.53.R3734
4. Cappellini G, Mancini M, Pagano G, Lombardi P, Livi L, Siciliani de Cumis M, Cancio P, Pizzocaro M, Calonico D, Levi F, Sias C, Catani J, Inguscio M, Fallani L (2014) Direct observation of coherent interorbital spin-exchange dynamics. Phys Rev Lett 113(120):402. https://doi.org/10.1103/PhysRevLett.113.120402
5. Cazalilla MA, Ho AF, Ueda M (2009) Ultracold gases of ytterbium: ferromagnetism and Mott states in an SU(6) Fermi system. New J Phys 11(10):103033. https://doi.org/10.1088/1367-2630/11/10/103033
6. Chanelière T, Meunier JL, Kaiser R, Miniatura C, Wilkowski D (2005) Extra-heating mechanism in Doppler cooling experiments. J Opt Soc Am B 22(9):1819–1828. https://doi.org/10.1364/JOSAB.22.001819
7. Cherng RW, Refael G, Demler E (2007) Superfluidity and magnetism in multicomponent ultracold fermions. Phys Rev Lett 99(13):130406. https://doi.org/10.1103/physrevlett.99.130406
8. Choi SK, Park SE, Chen J, Minogin VG (2008) Three-dimensional analysis of the magneto-optical trap for (1+3)-level atoms. Phys Rev A 77(1):015405. https://doi.org/10.1103/physreva.77.015405

9. Dalfovo F, Giorgini S, Pitaevskii LP, Stringari S (1999) Theory of Bose-Einstein condensation in trapped gases. Rev Mod Phys 71(3):463. https://doi.org/10.1103/RevModPhys.71.463
10. DeMarco B, Papp SB, Jin DS (2001) Pauli blocking of collisions in a quantum degenerate atomic Fermi gas. Phys Rev Lett 86(24):5409–5412. https://doi.org/10.1103/physrevlett.86.5409
11. Dörscher S, Thobe A, Hundt B, Kochanke A, Le Targat R, Windpassinger P, Becker C, Sengstock K (2013) Creation of quantum-degenerate gases of ytterbium in a compact 2D-/3D-magneto-optical trap setup. Rev Sci Instrum 84(4):043109. https://doi.org/10.1063/1.4802682
12. Dreon D, Sidorenkov LA, Bouazza C, Maineult W, Dalibard J, Nascimbene S (2017) Optical cooling and trapping of highly magnetic atoms: the benefits of a spontaneous spin polarization. J Phys B At Mol Opt Phys 50(6):065005. https://doi.org/10.1088/1361-6455/aa5db5
13. Drewsen M, Laurent P, Nadir A, Santarelli G, Clairon A, Castin Y, Grison D, Salomon C (1994) Investigation of sub-Doppler cooling effects in a cesium magneto-optical trap. Appl Phys B 59(3):283–298. https://doi.org/10.1007/BF01081396
14. Foot CJ (2004) Atomic physics. Oxford University Press
15. Fuchs J, Duffy GJ, Veeravalli G, Dyke P, Bartenstein M, Vale CJ, Hannaford P, Rowlands WJ (2007) Molecular Bose-Einstein condensation in a versatile low power crossed dipole trap. J Phys B At Mol Opt Phys 40(20):4109–4118. https://doi.org/10.1088/0953-4075/40/20/011
16. Fukuhara T, Sugawa S, Takahashi Y (2007a) Bose-Einstein condensation of an ytterbium isotope. Phys Rev A 76(5):051604. https://doi.org/10.1103/PhysRevA.76.051604
17. Fukuhara T, Takasu Y, Kumakura M, Takahashi Y (2007b) Degenerate Fermi gases of ytterbium. Phys Rev Lett 98(3):030401. https://doi.org/10.1103/PhysRevLett.98.030401
18. Gorshkov AV, Hermele M, Gurarie V, Xu C, Julienne PS, Ye J, Zoller P, Demler E, Lukin MD, Rey AM (2010) Two-orbital SU(N) magnetism with ultracold alkaline-earth atoms. Nat Phys 6(4):289–295. https://doi.org/10.1038/nphys1535
19. Grimm R, Weidemüller M, Ovchinnikov YB (2000) Optical dipole traps for neutral atoms. Adv At Mol Opt Phys 42:95–170. https://doi.org/10.1016/s1049-250x(08)60186-x
20. Hansen AH, Khramov AY, Dowd WH, Jamison AO, Plotkin-Swing B, Roy RJ, Gupta S (2013) Production of quantum-degenerate mixtures of ytterbium and lithium with controllable interspecies overlap. Phys Rev A 87(1):013615. https://doi.org/10.1103/PhysRevA.87.013615
21. Hermele M, Gurarie V, Rey AM (2009) Mott insulators of ultracold fermionic alkaline earth atoms: underconstrained magnetism and chiral spin liquid. Phys Rev Lett 103(13):135301. https://doi.org/10.1103/physrevlett.103.135301
22. Honerkamp C, Hofstetter W (2004) Ultracold fermions and the SU(N) Hubbard model. Phys Rev Lett 92(17):170403. https://doi.org/10.1103/physrevlett.92.170403
23. Hung CL, Zhang X, Gemelke N, Chin C (2008) Accelerating evaporative cooling of atoms into Bose-Einstein condensation in optical traps. Phys Rev A 78(1):011604. https://doi.org/10.1103/PhysRevA.78.011604
24. Johnson W, Kolb D, Huang KN (1983) Electric-dipole, quadrupole, and magnetic-dipole susceptibilities and shielding factors for closed-shell ions of the He, Ne, Ar, Ni (Cu$^+$), Kr, Pb, and Xe isoelectronic sequences. At Data Nucl Data Tables 28(2):333–340. https://doi.org/10.1016/0092-640x(83)90020-7
25. Katori H, Ido T, Isoya Y, Kuwata-Gonokami M (1999) Magneto-optical trapping and cooling of strontium atoms down to the photon recoil temperature. Phys Rev Lett 82:1116–1119. https://doi.org/10.1103/PhysRevLett.82.1116
26. Kemp SL, Butler KL, Freytag R, Hopkins SA, Hinds EA, Tarbutt MR, Cornish SL (2016) Production and characterization of a dual species magneto-optical trap of cesium and ytterbium. Rev Sci Instrum 87(2):023105. https://doi.org/10.1063/1.4941719
27. Kerman AJ, Vuletić V, Chin C, Chu S (2000) Beyond optical molasses: 3D Raman sideband cooling of atomic cesium to high phase-space density. Phys Rev Lett 84:439–442. https://doi.org/10.1103/PhysRevLett.84.439
28. Ketterle W (2002) Nobel lecture: when atoms behave as waves: Bose-Einstein condensation and the atom laser. Rev Mod Phys 74(4):1131–1151. https://doi.org/10.1103/RevModPhys.74.1131

29. Ketterle W, Druten NV (1996) Evaporative cooling of trapped atoms. Adv At Mol Opt Phy 37:181–236. https://doi.org/10.1016/S1049-250X(08)60101-9
30. Ketterle W, Zwierlein MW (2008) Making, probing and understanding ultracold Fermi gases. Riv Nuovo Cim 31:247–422. https://doi.org/10.1393/ncr/i2008-10033-1
31. Ketterle W, Durfee D, Stamper-Kurn D (1999) Making, probing and understanding Bose-Einstein condensates. IOS Press, pp 67–176. https://doi.org/10.3254/978-1-61499-225-7-67
32. Kitagawa M, Enomoto K, Kasa K, Takahashi Y, Ciuryło R, Naidon P, Julienne PS (2008) Two-color photoassociation spectroscopy of ytterbium atoms and the precise determinations of s-wave scattering lengths. Phys Rev A 77(1):012719. https://doi.org/10.1103/physreva.77.012719
33. Leggett AJ (2001) Bose-Einstein condensation in the alkali gases: some fundamental concepts. Rev Mod Phys 73(2):307. https://doi.org/10.1103/RevModPhys.73.307
34. Livi L, Cappellini G, Diem M, Franchi L, Clivati C, Frittelli M, Levi F, Calonico D, Catani J, Inguscio M, Fallani L (2016) Synthetic dimensions and spin-orbit coupling with an optical clock transition. Phys Rev Lett 117(22):220401. https://doi.org/10.1103/physrevlett.117.220401
35. Ludlow AD (2008) The strontium optical lattice clock: optical spectroscopy with sub-hertz accuracy. PhD thesis
36. Mancini M, Pagano G, Cappellini G, Livi L, Rider M, Catani J, Sias C, Zoller P, Inguscio M, Dalmonte M, Fallani L (2015) Observation of chiral edge states with neutral fermions in synthetic Hall ribbons. Science 349(6255):1510–1513. https://doi.org/10.1126/science.aaa8736
37. Maruyama R, Wynar RH, Romalis MV, Andalkar A, Swallows MD, Pearson CE, Fortson EN (2003) Investigation of sub-Doppler cooling in an ytterbium magneto-optical trap. Phys Rev A 68(011):403. https://doi.org/10.1103/PhysRevA.68.011403
38. Mitroy J, Safronova MS, Clark CW (2010) Theory and applications of atomic and ionic polarizabilities. J Phys B At Mol Opt Phys 43(20):202001. https://doi.org/10.1088/0953-4075/43/20/202001
39. Mukaiyama T, Katori H, Ido T, Li Y, Kuwata-Gonokami M (2003) Recoil-limited laser cooling of ^{87}Sr atoms near the Fermi temperature. Phys Rev Lett 90(11):113002. https://doi.org/10.1103/physrevlett.90.113002
40. Ozawa H, Taie S, Takasu Y, Takahashi Y (2018) Antiferromagnetic spin correlation of SU(N) Fermi gas in an optical dimerized Superlattice. Phys Rev Lett 121(22):225303 https://doi.org/10.1103/PhysRevLett.121.225303
41. Roy R, Green A, Bowler R, Gupta S (2016) Rapid cooling to quantum degeneracy in dynamically shaped atom traps. Phys Rev A 93(4):043403. https://doi.org/10.1103/PhysRevA.93.043403
42. Safronova MS, Arora B, Clark CW (2006) Frequency-dependent polarizabilities of alkali-metal atoms from ultraviolet through infrared spectral regions. Phys Rev A 73(2):022505. https://doi.org/10.1103/PhysRevA.73.022505
43. Safronova MS, Safronova UI, Clark CW (2016) Magic wavelengths, matrix elements, polarizabilities, and lifetimes of Cs. Phys Rev A 94(1):012505. https://doi.org/10.1103/physreva.94.012505
44. Scazza F (2015) Probing SU(N)-symmetric orbital interactions with ytterbium Fermi gases in optical lattices. PhD thesis
45. Scazza F, Hofrichter C, Hofer M, De Groot PC, Bloch I, Folling S (2014) Observation of two-orbital spin-exchange interactions with ultracold SU(N)-symmetric fermions. Nat Phys 10(10):779–784. https://doi.org/10.1038/nphys3061
46. Stellmer S, Tey MK, Huang B, Grimm R, Schreck F (2009) Bose-Einstein condensation of strontium. Phys Rev Lett 103(20):200401. https://doi.org/10.1103/PhysRevLett.103.200401
47. Sugawa S, Yamazaki R, Taie S, Takahashi Y (2011) Bose-Einstein condensate in gases of rare atomic species. Phys Rev A 84(1):011610. https://doi.org/10.1103/PhysRevA.84.011610
48. Taie S, Takasu Y, Sugawa S, Yamazaki R, Tsujimoto T, Murakami R, Takahashi Y (2010) Realization of a SU (2)× SU (6) system of fermions in a cold atomic gas. Phys Rev Lett 105(19):190401. https://doi.org/10.1103/PhysRevLett.105.190401

49. Taie S, Yamazaki R, Sugawa S, Takahashi Y (2012) An SU(6) Mott insulator of an atomic Fermi gas realized by large-spin Pomeranchuk cooling. Nat Phys 8(11):825–830. https://doi. org/10.1038/nphys2430
50. Takasu Y, Takahashi Y (2009) Quantum degenerate gases of ytterbium atoms. J Phys Soc Jpn 78(1):012001. https://doi.org/10.1143/JPSJ.78.012001
51. Takasu Y, Maki K, Komori K, Takano T, Honda K, Kumakura M, Yabuzaki T, Takahashi Y (2003) Spin-singlet Bose-Einstein condensation of two-electron atoms. Phys Rev Lett 91(4):040404. https://doi.org/10.1103/PhysRevLett.91.040404
52. Vaidya VD, Tiamsuphat J, Rolston SL, Porto JV (2015) Degenerate Bose-Fermi mixtures of rubidium and ytterbium. Phys Rev A 92(043):604. https://doi.org/10.1103/PhysRevA.92. 043604
53. Weber T, Herbig J, Mark M, Nägerl HC, Grimm R (2003) Three-body recombination at large scattering lengths in an ultracold atomic gas. Phys Rev Lett 91(123):201. https://doi.org/10. 1103/PhysRevLett.91.123201
54. Weiner J, Bagnato VS, Zilio S, Julienne PS (1999) Experiments and theory in cold and ultracold collisions. Rev Mod Phys 71(1):1–85. https://doi.org/10.1103/revmodphys.71.1
55. Xu C (2010) Liquids in multiorbital SU(N) magnets made up of ultracold alkaline-earth atoms. Phys Rev B 81(14):144431. https://doi.org/10.1103/physrevb.81.144431
56. Xu X, Loftus TH, Hall JL, Gallagher A, Ye J (2003) Cooling and trapping of atomic strontium. J Opt Soc Am B 20(5):968–976. https://doi.org/10.1364/josab.20.000968

Chapter 5
A Quantum Degenerate Gas of Cs

The efficient preparation of a high PSD mixture in a dipole trap is an essential step towards molecule production. With the dipole trap carefully tailored to optimise the production of degenerate gases of Yb atoms, we then concentrated on the production of a Cs BEC. Although Cs possesses favourable properties for laser cooling [5, 21] due to its large mass and hyperfine splitting, the production of a Cs BEC is a challenging task. This is due to its troublesome scattering properties [3, 17, 34, 42] which prevented groups from achieving a Cs BEC until 2002 [41], some 7 years after the first BECs in alkali atoms were observed [1, 8]. Even after the first observation of Cs BEC only a handful of other groups worldwide have reported production of Cs BECs [14, 19, 29, 33].

In this chapter we describe the implementation of degenerate Raman sideband cooling (DRSC) as an efficient method for cooling and polarising Cs atoms. We outline the setup of an additional dipole trap for Cs and described the necessary techniques for cooling Cs to quantum degeneracy.

5.1 Cs MOT and Sub-doppler Cooling

The initial stage in the preparation of a Cs BEC is the collection of Cs atoms in a MOT. The detunings of the Cs beams are the same as those given in Table 3.2 in Sect. 3.3.1. To load the Cs MOT we apply a magnetic field gradient of 7.7 G/cm, MOT repump light with intensity 7 mW/cm^2 and use a total MOT beam intensity of 80 mW/cm^2. After 6 s of MOT loading we accumulate 3×10^8 Cs atoms in the MOT.

After the initial loading of the Cs MOT, the ZS coils and light are switched off and the Cs atoms are compressed using a compressed MOT stage. During the compressed MOT stage the quadrupole gradient is increased to 14 G/cm, the repump intensity is decreased to 25 μW/cm^2 and the MOT detuning is increased to -20 MHz. This compresses the MOT by reducing the radiation pressure from the scattered photons

© Springer Nature Switzerland AG 2019   91
A. Guttridge, *Photoassociation of Ultracold CsYb Molecules
and Determination of Interspecies Scattering Lengths*, Springer Theses,
https://doi.org/10.1007/978-3-030-21201-8_5

and increases the MOT density. Following the compression, the shim coils are used to manoeuvre the MOT to the centre of a set of Raman lattice beams used for degenerate Raman sideband cooling.

After the MOT has been translated into the correct position for overlap with the Raman lattice beams, we perform sub-Doppler cooling using an optical molasses stage [27]. This is performed by turning off the quadrupole gradient and increasing the MOT detuning to -45 MHz. The optical molasses stage uses polarisation gradient cooling [7] to reach temperatures well below the Doppler limit, which is 125 μk for Cs [36]. It is critical during this stage to null any residual magnetic fields in the chamber using the shims, as the presence of a magnetic field can cause Larmor precession and Zeeman shifts of the magnetic sublevels. These effects compete with the optical pumping and AC stark shifts and interfere with the cooling [27]. Following the 25 ms long optical molasses stage the Cs atoms are cooled to 20 μK.

5.2   Degenerate Raman Sideband Cooling of Cs

The most prominent method for precooling an atomic ensemble for the preparation of a Cs BEC is degenerate Raman sideband cooling (DRSC). DRSC is a powerful technique that allows the production of $>10^7$ atoms at temperatures around 1 μK in the absolute ground state $|F = 3, m_F + 3\rangle$. The pioneering work on DRSC of Cs [15, 22, 38, 39] was an essential step towards the achievement of the first Cs BEC [42]. DRSC is an attractive technique as the cooling mechanism is quick, efficient and, crucially, produces Cs in the only state in which it has been condensed, $|F = 3, m_F = +3\rangle$. The use of the absolute ground state mitigates the large 2-body loss rate that hindered other BEC attempts working with the magnetically trappable $|4, +4\rangle$ [2, 3, 34] and $|3, -3\rangle$ states [12, 17, 37]. As we saw in Sect. 2.2.1, the $|3, +3\rangle$ state has a rich Feshbach structure which allows the scattering length to be tuned near to an Efimov minimum in the 3-body loss rate [42], allowing efficient evaporative cooling of Cs in an optical trap.

5.2.1   Degenerate Raman Sideband Cooling

The DRSC mechanism is displayed in Fig. 5.1a. The atoms initially in the $F = 3, m_F = +3$ state occupy a high-lying vibrational level v of the Raman lattice. By applying a small magnetic field the magnetic sublevels are Zeeman shifted such that the $|F = 3, m_F = +1, v - 2\rangle, |F = 3, m_F = +2, v - 1\rangle$ and $|F = 3, m_F = +3, v\rangle$ states are degenerate. Coupling is induced between the states via two photon Raman transitions which are driven by the lattice beams, in a similar setup to the original Cs implementations of DRSC [22, 38]. The atoms in $|F = 3, m_F = +1, v - 2\rangle$ are then recycled back into the $F = 3, m_F = +3$ state by optical pumping using the $6S_{1/2}, F = 3 \rightarrow 6P_{3/2}, F' = 2$ transition. For the cooling scheme to operate effectively it is essential that the optical confinement is strong enough that the excitation occurs in the Lamb-Dicke regime, to prevent the atom changing vibrational state by the scattered lattice light. If $\eta = \sqrt{E_{\text{Rec}}/\hbar\omega} < 1$ is satisfied, two quanta of vibrational energy are removed for each cycle of the cooling. The optical pumping light, called the polariser, drives σ^+ transitions to the $F' = 2$ state which predominantly decays to $m_F = 3$. The cooling scheme proceeds quickly, facilitated by the fast Raman coupling, until the atom reaches the $|F = 3, m_F = +3, v = 1\rangle$ state. The atom is then transferred into the $|F = 3, m_F = +2, v = 0\rangle$ state but cannot access the $m_F = 1$ sublevel through a Raman transition. A weak π component of the polariser light slowly pumps the atoms from $|F = 3, m_F = +2, v = 0\rangle$ into the dark $|F = 3, m_F = +3, v = 0\rangle$ state via the $F' = 2$ state.

Figure 5.1b shows the optical setup for DRSC in our experiment. The Raman lattice is operated in a four-beam configuration, with one horizontal standing wave (retro-reflected) beam, one horizontal running wave beam and one vertical running wave beam. In this configuration, phase fluctuations in the lattice beams cause global translations of the entire lattice, while the depth and shape of the lattice sites are unaffected [11]. The three near-orthogonal linearly polarised lattice beams are resonant with the $F = 4 \rightarrow F' = 4$ transition, detuned by -9.2 GHz from the atoms in the $F = 3$ state. The choice of this transition for the lattice means that the lattice beams provide repumping from the metastable $F = 4$ ground state into the $F = 3$ manifold, in addition to providing the necessary Raman coupling between the magnetic sublevels. The polariser beam operates on the $F = 3 \rightarrow F' = 2$ transition and is

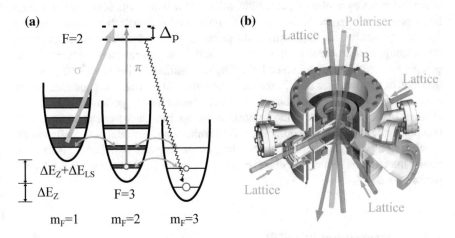

Fig. 5.1 Degenerate Raman sideband cooling setup and implementation. **a** Diagram illustrating the degenerate Raman sideband cooling process. When a magnetic field is applied such that the vibrational levels of the sublevels are degenerate, 2 vibrational quanta of energy are removed per cycle via Raman transitions between these levels. The fast σ^+ optical pumping transfers the population to the dark $m_F = 3$ state, where the atoms accumulate in the lowest vibrational level. **b** Optical setup for DRSC. The polariser beam is circularly polarised and at a small angle to the quantisation axis provided by the magnetic field, B

circularly polarised so as to drive σ^+ transitions to the $m_F = +3$ state. The weak π component is provided by tilting the magnetic field by a small angle with respect to the polariser beam.

Experimentally, the lattice light is delivered to the optical table by two optical fibres, carrying light for the horizontal and vertical lattice beams. The lattice beams have a $1/e^2$ beam diameter of 4 mm and the total available power for the lattice is 100 mW. In the initial search for DRSC we used smaller lattice beams of 2 mm diameter to increase the depth and trap frequencies of the lattice which allowed easier observation of DRSC. However, after optimisation we found increasing the size of the lattice beams yielded a larger number of captured atoms at the same temperature, likely due to the increased trap volume. Control of the total power in the lattice is provided by a 120 MHz AOM before the optical fibres. Control of the power distribution between the horizontal and vertical lattice beams is provided by a $\lambda/2$ waveplate before the PBS used to split the light into the two fibres. After the output of the horizontal lattice fibre, the lattice beams are split into standing and running wave beams by a PBS. The polarisation of the vertical and horizontal beams are set by independent $\lambda/2$ waveplates. We note that some experiments [9, 38, 39] have found that using a $\lambda/4$ waveplate before the standing wave retro-reflecting mirror improved the DRSC performance, however, we did not observe any improvement in our experiment.

The polariser light is generated by the repump laser and is delivered to the experiment using a polarisation-maintaining fibre. Like the lattice beams, the polariser beam has a $1/e^2$ beam diameter of 4 mm. The detuning from the $F = 3 \rightarrow F' = 2$ transition is set by a double passed 200 MHz AOM before the fibre. The polariser beam propagates vertically through the chamber and is circularly polarised using a $\lambda/4$ waveplate after the final mirror. The polariser propagates at a small angle to the quantisation axis provided by the magnetic field. This prevents any ellipticity in the polarisation creating σ^- polarised light. Slightly shifting the field from the propagation axis of the polariser transforms this ellipticity into a small π component, which is necessary to pump out of the $m_F = +2$ dark state for σ^+ polarised light. Small changes in the polarisation of the polariser beam can introduce heating through a σ^- component or a too large π component. We eliminate any residual polarisation drifts by placing a polarising cube after the fibre, converting any polarisation drifts into power fluctuations, which we observe to have a much weaker effect on the DRSC performance.

5.2.2   Observation of DRSC

The high scattering rate of the near-resonant lattice light necessitates a significant cooling rate for the observation of atoms trapped in the lattice. In the near detuned lattice the heating rate from scattered light is $\dot{T} = 500\,\mu K/s$. This heating rate is fairly large, however, in one cooling cycle we remove $2\,\hbar\omega \simeq k_B \times 5\,\mu K$ from the atoms. The speed of the cooling cycle is limited by optical pumping, which occurs at a rate $\Gamma_{OP} \simeq 300$ kHz, depending on the polariser parameters. This results in a large cooling rate of 1.5 K/s which dominates over the heating rate.

However, to obtain this large cooling rate a large number of DRSC parameters must be near optimal. We used a range of independent diagnostics to test various parameters and reduce the size of the parameter space in which we searched for evidence of DRSC.

Lattice polarisation: To be able to trap atoms in the lattice it was important to simulate the trap formed by the lattice beams and identify the optimal polarisations for the lattice beams. This is crucial in our experimental setup because optical access dictates that the lattice beams are not orthogonal, unlike in ideal implementations of DRSC. The simulation allows the trap depth and trap frequencies of the lattice to be maximised. Larger trap frequencies allow a larger quantisation field (which is typically small: of the order of a few 100 mG) and are also important for realising the Lamb-Dicke regime.

Magnetic fields: For the weak magnetic fields required to shift the vibrational levels into degeneracy, we first experimentally determined the residual magnetic field in the science chamber using a triple axis magnetometer (Honeywell HMC5883L magnetic sensor). The magnetic fields in the horizontal plane were nulled using the magnetometer, an essential requirement as the weak magnetic quantisation axis must not be tilted by too large an angle from the direction of the polariser. The tilting of the field can introduce a large π or σ^- component in the polariser, causing heating.

Polariser: The frequency, polarisation and alignment of the polariser beam may be observed using magnetic recapture measurements. The number of Cs atoms captured in the magnetic trap[1] is reduced when the polariser beam drives σ^+ transitions to the anti-trapped $|F = 3, m_F = +3\rangle$ state. The polarisation, alignment and magnetic bias field for the polariser beam was optimised by reducing the number of recaptured atoms. The $F = 3 \rightarrow F' = 2$ resonance frequency was also checked using this technique but the polariser beam should be set to a moderate blue detuning when searching for DRSC.

Lattice alignment: The lattice beams were aligned onto the MOT position after molasses by imaging the atoms without repump light, after a short pulse of lattice light [9]. The lattice light quickly pumps the atoms into the $F = 3$ state, which is dark to the imaging light in the absence of repump light. Therefore, absorption images of the Cs atoms after the lattice light pulse show an absence of Cs atoms where the lattice light is incident on the cloud and allows alignment to the centre of the cloud. The diameter of the lattice beams were apertured down to a few hundred μm for fine adjustment of the alignment. The polariser beam was aligned using a similar technique.

With the lattice beams aligned onto the atoms and the polarisations of the lattice beams set to produce a sufficient trap depth to capture the Cs atoms, we began to look

---

[1]The magnetic trap is formed by the quadrupole coils in the absence of light.

for evidence of DRSC in our setup. The routine for observation of DRSC is as follows. We initially prepare the Cs atoms by loading a MOT, before compressing the MOT via gradient, detuning and intensity ramps. Once the MOT has been compressed, the atoms are further cooled using optical molasses. All the parameters for these steps are as described in the previous section. Following the molasses stage, the atoms have a low enough temperature to be trapped in the lattice. However, before DRSC can be performed the atoms are required to populate the $F = 3$ ground state. The atoms are depumped into the $F = 3$ ground state by extinguishing the MOT repump light 500 μs before the MOT beams are shuttered off, this step immediately follows the 25 ms molasses stage. The shim coils are used to set a vertical magnetic field of $\simeq 300$ mG and the Raman lattice is ramped up to full power in 600 μs. The polariser beam is then switched on for a variable time in which the cooling occurs. To avoid heating during release of the sample, the atoms are released into free space by linearly ramping down the lattice power over 1.2 ms, after which the light is turned off completely. The adiabatic release of the atoms is critical for achieving low free-space temperatures after the application of DRSC [22]. The final temperature in our setup may be improved by the implementation of a slightly more complex ramp that maximises adiabaticity [21].

Initially, it is possible to observe confinement of atoms in the lattice in the absence of polariser light. The lifetime of atoms in the near-resonant lattice is of order $\simeq 20$ ms, so observation of this confinement requires a short lattice hold time. The remaining atoms are also significantly heated by the lattice light, requiring a short time of flight after release from the lattice. Therefore, we use a time which is just long enough for the unconfined MOT atoms to drop away from the lattice region. Evidence of Cs atoms trapped in the near-resonant Raman lattice is shown in Fig. 5.2a, where the atoms are held in the lattice for 25 ms and then imaged after a time of flight of 15 ms. We found that it was critical to observe Cs atoms effected by the lattice as in Fig. 5.2a before moving onto scanning other parameters. Figure 5.2b shows the Cs atoms under the same conditions but with the application of the polariser. In the presence of the polariser beam the Cs atoms are rapidly cooled by DRSC and remain confined in the lattice.

We verified the quantum state of the cooled Cs atoms after release from the lattice using a Stern-Gerlach experiment, the results of which are shown in Fig. 5.2c. The left hand image shows the spin state populations when the polariser beam is shuttered off before the Raman lattice is extinguished. We observe Cs atoms distributed over the $m_F = +1, +2$ and $+3$ states due to pumping by the Raman lattice. The right hand image in Fig. 5.2c shows the spin state populations with the polariser present during the Raman lattice ramp off step. The atoms remain well polarised using this routine, we observe $>90\%$ of the Cs atoms populate the $|F = 3, m_F = +3\rangle$ state. In this experiment, the atomic polarisation is preserved during the rise time of the MOT and Bias coils by using a vertical magnetic bias field provided by the vertical shim coil. The shim bias field is applied until the Bias coil is ramped to its final value used for the levitated absorption imaging.

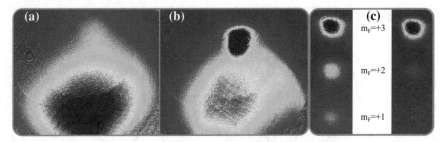

Fig. 5.2 Observation of DRSC. **a** and **b**: Absorption images of Cs after 25 ms in the Raman lattice and with 15 ms TOF. In **a** the polariser is not present but there is still some confinement of Cs atoms in the lattice. In **b** the polariser beam is present and a large fraction of atoms are trapped in the lattice. **c** Stern-Gerlach experiment to resolve the population of the Cs magnetic sublevels. The left image shows the spin state distribution when the polariser is off during the lattice off ramp. The right image shows the result when the polariser remains on during the lattice off ramp. When the polariser remains on >90% of the Cs atoms populate the $F = 3, m_F = +3$ state

5.2.3 Optimisation of DRSC

Once evidence of DRSC was obtained, we began to optimise the wealth of parameters which influence the performance of DRSC. The figure of merit we used to optimise the parameters was the number of atoms that remain trapped in the lattice. In Fig. 5.3 we present the optimisation of the magnetic field during DRSC. Following DRSC, the atoms are absorption imaged after a levitated time of flight, where we resolve the $m_F = +1, +2, +3$ populations. The number plotted in the figure is that of the $m_F = +3$ state. The optimal fields in the horizontal plane are around the zero field value measured with the magnetic field probe. The discrepancy in the East-West direction may be due to the required tilt of the field with respect to the polariser. Or, the small offset could be explained by the fact the magnetic field measurements were not in situ and the real field zero is slightly shifted from the extrapolated zero field position measured by the field probe. In the vertical direction the number of atoms captured rises linearly as the Zeeman shift of the magnetic sub states approaches the vibrational spacing. The lattice simulations predicted a value for the vertical trap frequency of $2\pi \times 70$ kHz, which corresponds to a magnetic field of 200 mG, in good agreement with the results. For fields >200 mG the trapped atom number decays slowly and we do not resolve the individual vibration frequencies of the lattice sites. The lack of well-defined vibration frequencies may be due to the spatial variation of trapping frequencies across the cloud [22]. In other experiments using DRSC [13] groups have found a significant improvement in the cooling performance by ramping the vertical magnetic field during the DRSC stage. The magnetic field ramp addresses lattice sites near the edge of the lattice with different vibrational frequencies to those at the centre. The different vibrational frequencies occur due to the finite size of the lattice beams. We find the step only makes a slight improvement to the DRSC performance.

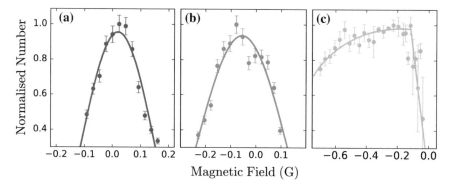

Fig. 5.3 Optimisation of magnetic fields for DRSC. Normalised number of Cs atoms in the $F = 3, m_F = +3$ state captured by DRSC cooling as a function of magnetic field in the laboratory frame. **a** Magnetic fields in the North-South direction. **b** Magnetic fields in the East-West direction. **c** Magnetic fields in the vertical direction. Zero magnetic field is determined by measurements with a triple axis magnetometer. Solid lines are fits to guide the eye

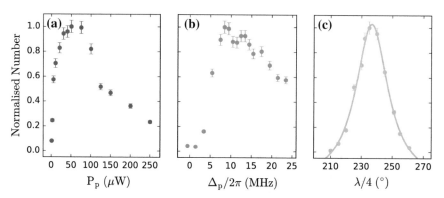

Fig. 5.4 Optimisation of polariser parameters for DRSC. Normalised number of Cs atoms in the $F = 3, m_F = +3$ state captured by DRSC cooling as a function of various polariser parameters. **a** Number of Cs atoms versus polariser power. **b** Number of Cs atoms versus polariser detuning, there is a clear minimum at zero detuning. **c** Number of Cs atoms versus angle of the polariser $\lambda/4$ waveplate. The solid line is a Lorentzian fit to the data

Figure 5.4 shows the optimisation of the polariser parameters. The dependence of the DRSC performance on the power of the polariser is shown in Fig. 5.4a. Sufficient power is required to pump atoms out of the possible $|m_F = +2, v = 0\rangle$ σ^+ dark state using a small π component of the polariser. For larger powers we observe a decrease in the atom number, this is also evident for longer DRSC times in Fig. 5.5. As the polariser is a running wave beam, the light exacts a pushing force on atoms, which may push them out of the lattice causing the observed loss. Another explanation for the decay may be the population of higher lattice bands by the atoms. The population of higher lattice bands allow atoms to tunnel more easily to other lattice sites causing loss [10].

The number of atoms cooled by DRSC is also strongly dependent on the detuning of the polariser light from the $F = 3 \rightarrow F' = 2$ transition. The maximum number of atoms are trapped for blue detunings between 8–12 MHz but almost zero atoms are trapped for a resonant polariser beam. This is due to the light shift, ΔE_{LS}, of the $m_F = 1$ level by the strong σ^+ polariser. For blue detuning $m_F = 1$ is shifted upwards (see Fig. 5.1a) in energy and the cooling performs optimally with only a small dependence on the magnitude of the detuning. However, when the polariser is near resonant or red detuned, the $m_F = 1$ level is shifted below $m_F = 2$ level and the cooling turns into a heating process, causing loss from the lattice.

Figure 5.4c shows the dependence of the atom number on the polarisation of the polariser beam. The cooling performance works optimally when the strong heating σ^- component is minimised. The angle of the $\lambda/4$ was first set by using the polariser beam as an optical pumping beam for magnetic recapture. This verified the magnetic field measurements by performing optical pumping with a small bias field. Then, once DRSC was observed and the magnetic fields were optimised, the waveplate angle was optimised by maximising the Cs atom number.

Once all these parameters were optimised, we investigated the temporal dependence of the cooling scheme. After the release from the lattice, the atoms in $|F = 3, m_F = +3\rangle$ are magnetically levitated for 60–80 ms by a 31.3 G/cm magnetic field gradient and subsequently imaged using absorption imaging. The temperature of the atoms was measured using time of flight expansion. Care must be taken when using when measuring the temperature using the levitated TOF technique as the horizontal width of the cloud is effected by the combined curvature of the MOT and Bias coils. Therefore, only the vertical expansion is used when measuring the temperature. Figure 5.5 shows the dependence of the temperature and number on the length of the DRSC cooling stage. The cooling occurs rapidly, with the atoms cooled to a few μK in 5 ms. We observe a decay in the number of atoms for longer DRSC times which we attribute to be the same effect(s) that caused the loss of atoms in the polariser power measurements. As a result of these measurements we typically apply DRSC for 8 ms. The fully optimised DRSC routine uses 4 mm diameter lattice beams (2 mm beams were used in Fig. 5.5), allowing the production of 6×10^7 atoms

Fig. 5.5 Number of atoms (solid circles) and temperature (open circles) after DRSC as a function of cooling time. The dashed line is an exponential decay fitted to the data

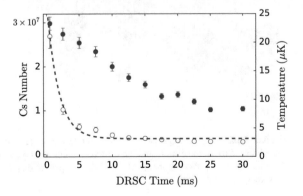

polarised in the $|F = 3, m_F = +3\rangle$ state at a temperature of $1.7\,\mu$K. This is a good starting point for the evaporation of Cs in an optical trap.

5.3   Optical Trapping of Cs

After the application of DRSC, $>90\%$ of the Cs atoms populate the $|F = 3, m_F = +3\rangle$ state. To enable evaporative cooling, the Cs atoms must be trapped optically, as they populate a high-field seeking state and cannot be trapped magnetically. Transfer of this sample into the dimple trap (as used for Yb) is very inefficient due to the poor mode matching between the dimple trap (cloud diameter $\sim$100 μm) and the sample in the Raman lattice (cloud diameter $\sim$3 mm). When the power in the dimple trap beams are ramped up, the atoms away from the centre of the beams gain a large amount of potential energy. The potential energy gained is then converted into kinetic energy and heats the sample causing poor transfer of the atoms into the trap.

5.3.1   Reservoir Trap

To mitigate the mode matching problem we installed a large volume reservoir trap that is used to transfer the DRSC cooled atoms into the tightly-confining dimple trap. The two optical traps used for Cs are shown in Fig. 5.6. The reservoir trapping light is produced by a 50 W IPG laser operating at 1070 nm. The reservoir trap is composed

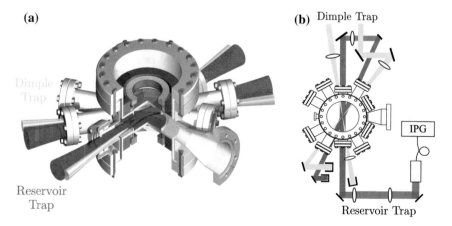

Fig. 5.6   Optical dipole traps used for Cs. **a** Rendering of the science chamber and the two crossed optical dipole traps, the dimple and the reservoir. **b** Optical layout of the reservoir and dimple traps. The additional optical components of the dimple trap were displayed previously in Fig. 4.3 and are not shown here for clarity

of two beams crossed at 25° with waists of 640 μm and 440 μm respectively. This crossing angle is small and far from the 90° angle desired but we are constrained to this geometry due to the lack of usable viewports. The trap is operated in a bow-tie configuration which allows the recycling of beam power and facilitates the use of larger beam waists. The desired trap depth of 20 μK is produced using 20 W of optical power. The trap is aligned to the position of the Cs atoms following DRSC. Fine adjustment of the alignment may be made by minimising the amplitude of the COM oscillations of the Cs atoms after loading into the trap.

Due to the low trap depth, the trap is completely dominated by gravity in the absence of a magnetic field gradient. This is illustrated by the red calculated potential in Fig. 5.7a. For 20 W of power there is no confinement of Cs atoms in the vertical z direction due to the strong effect of gravity. To counteract the effect of gravity we employ a magnetic field gradient generated by the quadrupole coils. This generates an additional magnetic potential of $U_{Mag,z} = -m_F g_f \mu_B \frac{\partial B}{\partial z} z$ which for the high-field seeking $|F = 3, m_F = +3\rangle$ Cs atoms counteracts the gravitational potential. The gravitational potential is completely cancelled for a field gradient of 31.3 G/cm as shown by the blue potential in Fig. 5.7a.

Unfortunately, the application of a magnetic field gradient to counteract the effect of gravity also produces some anti-trapping along the axial direction of the trap. This is required by Gauss's law for magnetism (Maxwell II) $\nabla \cdot B = 0$, which dictates that a vertical magnetic field gradient also produces a field gradient radially in x and y. The magnetic potential in the horizontal plane $U_{Mag,H}$ is [16, 28]

$$U_{mag,H}(x, y) = -\frac{3}{4} \mu_B B_{bias} - \frac{1}{6} \frac{m^2 g^2}{\mu_B B_{bias}} \left(x^2 + y^2 \right), \tag{5.1}$$

where B_{bias} is the magnetic bias field applied in the z direction. This can be recast in the form of an inverse harmonic trapping potential

$$U_{mag,H}(x, y) = U_{bias} - \frac{1}{2} m \omega_{anti}^2 \left(x^2 + y^2 \right), \tag{5.2}$$

with anti-trapping frequency $\omega_{anti} = g \sqrt{\frac{m}{3 \mu_B B_{bias}}}$. The effect of anti-trapping in the axial direction for a small bias field is shown by the grey dotted line in Fig. 5.7b. It is evident from Eq. 5.2 that the application of a large bias field will negate the effect of the anti-trapping in the vertical direction by shifting the magnetic field zero further from the atoms, thereby reducing the curvature of the magnetic field at the atoms. The reduction of the anti-trapping with the application of larger bias fields is shown by the purple and pink lines in Fig. 5.7b. However, it is important to note that for Cs in the $|3, +3\rangle$ state, applying a larger bias field typically yields a large scattering length. For the 70 and 100 G fields used in the potentials shown by the purple and pink lines, the corresponding scattering lengths are $a_{Cs} \simeq 1300\, a_0$ and $1500\, a_0$. These values of the scattering length are very large but fortunately the initial density in the trap is not. Therefore, the higher scattering lengths may aid the initial

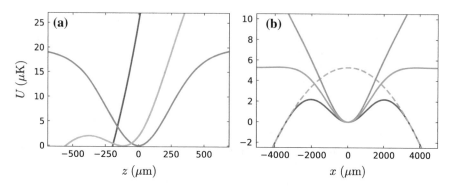

Fig. 5.7 Effect of magnetic fields on the reservoir trap. Simulated optical potentials for the reservoir trap using Cs atoms in the $|F = 3, m_F = +3\rangle$ state. For illustrative purposes the potential minimum has been set to zero for traps with a finite trap depth. **a** Trapping potential in the vertical direction for magnetic field gradients 0 G/cm (red), 24 G/cm (green) and 31 G/cm (blue). **b** Trapping potential in the axial direction for a levitating magnetic field gradient of 31 G/cm and magnetic bias fields 5 G (grey dashed), 70 G (purple), 100 G (pink) and 140 G (orange)

loading of the trap up to a point where the atomic density becomes too large and the scattering length and hence the bias field must be reduced.

In addition to the reduction of anti-trapping by moving the field zero away from the atoms, the Bias coils also provide a small trapping force due to their slight deviation from a Helmholtz configuration. Therefore, large bias fields can not only remove the anti-trapping effect of the gradient but create a 2D magnetic reservoir trap in the horizontal plane [18]. This effect is shown by the orange line in Fig. 5.7b. Unfortunately, it was not possible to reach these fields with the current setup of the Bias coils, so for loading the trap we use a bias field of 70 G. The calculated potential for this field is shown by the purple line in Fig. 5.7b. The simulation of the trapping potential shows that this bias field only partially reduces the anti-trapping and does not cancel the effect completely.

To transfer the DRSC atoms into the levitated trap we switch on the reservoir trapping light during the lattice off step. Immediately after the DRSC lattice beams are extinguished we apply a vertical bias field of 7 G using the vertical shim. This serves to preserve the atomic polarisation and shift the field zero below the atoms before the levitation gradient is applied. We then begin to ramp the Bias coil to its final value of 70 G and switch on the quadrupole gradient to levitate the atoms. Due to the finite rise time of the quadrupole coil we initially over-levitate the atoms to cancel the velocity acquired from their brief free-fall. After some empirical adjustments, the magnetic field gradient is ramped up to 52 G/cm in 3 ms and then ramped down to the final levitation gradient of 31.3 G/cm in 4 ms. Further improvement of the initial loading is performed by optimising the MOT compression step before DRSC. The MOT gradient turned out to be a particularly important parameter, and was varied to improve the mode matching in the MOT to DRSC lattice to reservoir trap transfer.

Using this routine we are able to load 2.5×10^7 atoms at 4 μK into the reservoir trap. However, after 2 s of plain evaporation, as the atoms equilibrate with the reservoir trap, this number reduces to 4×10^6 at 1.5 μK.

5.3.2   Dimple Trap

Evaporation to quantum degeneracy may be performed in the reservoir trap [18]. Evaporation by tilting the reservoir trap using the magnetic field gradient is very efficient as it maintains the initial trapping frequencies throughout the evaporation. However, for dual-species experiments and molecule production we require a common trapping potential for both Cs and Yb. Due to the absence of a magnetic moment in Yb, the Yb atoms cannot be trapped in the reservoir because it cannot be levitated with a magnetic field gradient. With this in mind we decided to produce a Cs BEC using the tried and tested 'dimple trick' [24, 32, 35, 40], the method used in the first Cs BEC experiments.

The 'dimple trick' (illustrated in Fig. 5.8) facilitates transfer of the atoms into a trapping potential with larger trapping frequencies for efficient evaporative cooling. The 33 μm and 72 μm dimple beams are ramped up to 80 mW and 200 mW respectively in 150 ms, forming a small dimple within the reservoir trapping potential. The much lower powers required to trap Cs at this wavelength meant we only operated the IPG laser at 20% of its maximum power (as opposed to 85% for Yb), corresponding to a power of 20 W out of the laser head. During the dimple power ramp the reservoir atoms begin to load into the dimple via elastic collisions. The elastic collision rate at this stage is fairly high as the bias field remains at 70 G which corresponds to a scattering length of $a_{\mathrm{Cs}} \simeq 1300\,a_0$. Although the density of the atoms in the dimple is increased appreciably, there is negligible heating to these atoms as they are cooled through collisions with the bath of reservoir atoms. Following a 200 ms hold step we

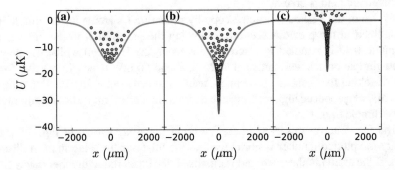

Fig. 5.8 The Cs 'dimple trick'. **a** Calculated trapping potential of the Cs reservoir trap. **b** Combined trapping potential of the reservoir (blue) and dimple (red) traps. **c** Removing the reservoir confinement leads to evaporation of the reservoir atoms (blue) and leaves a dense, cold sample of atoms confined solely by the dimple trap (red)

reduce the bias field to 25 G corresponding to a scattering length of $a_{Cs} \simeq 350\,a_0$; this is in order to reduce the three-body loss rate for the next stages where the density will be increased. We then remove the untrapped reservoir atoms by ramping the magnetic field gradient down to 24 G/cm in 400 ms, after which the reservoir beams are extinguished. The gradient ramp serves to evaporatively cool the reservoir atoms and enhance the loading of the dimple, during this step we also reduce the bias field to 22 G. After this stage we have 5×10^5 Cs atoms trapped in the dimple with a temperature of 1 μK.

Over the course of the dimple trick the PSD of the sample has increased by more than 3 orders of magnitude in contrast to a simple adiabatic compression.

5.4   Cs₂ Feshbach Resonances

The high density sample of Cs atoms trapped in the dimple is a good starting point for investigation of Feshbach resonances. The reason for the investigation of Cs₂ Feshbach resonances is twofold. Firstly, investigation of CsYb Feshbach resonances is one of the key objectives of the experiment, so development of techniques to observe Feshbach resonances using a simpler, single-species sample is useful. Secondly, Cs has a rich Feshbach structure with many low field resonances [4, 6] that we can use to calibrate our magnetic field. This is necessary as we currently do not have the capability to drive RF or microwave transitions between Cs internal states which is the most commonly used method for precise magnetic field calibration.

The rich Feshbach structure of the Cs $|3, +3\rangle$ state is shown in 5.9a for low magnetic fields. Even in this low-field region Cs possesses narrow and broad Feshbach resonances which offer precise control of the scattering length and thus the atomic properties. The near threshold bound state that produces the broad -12 G Feshbach resonance gives Cs its large zero field scattering length $a_{bg} = -2400\,a_0$. Cs clearly cannot be condensed at zero field so knowledge of the scattering length and therefore the magnetic field is crucial.

We begin the Feshbach spectroscopy by preparing Cs atoms in the dimple trap as outlined in the previous section. Following the preparation in the dimple, the magnetic field is scanned in discrete steps using the Bias coil whilst the powers in the dimple beams were reduced to 25 mW and 50 mW over 750 ms. We found that searching for Feshbach resonances using a forced evaporative cooling ramp to $T = 500$ nK produced higher contrast features compared to simply holding the atoms in the dimple trap.

The five features observed in the 14–60 G range are shown in Fig. 5.9b. For clarity the plotted number is normalised to the off-resonant value in each discrete scan. If the atom number were not normalised, the background number (away from a Feshbach resonance) varies by over a factor of four. This is because the three-body recombination loss rate scales $\propto a^4$ [42], so for magnetic fields far from the 22 G Efimov minimum in the three-body loss rate the atom loss is large. Near a Feshbach

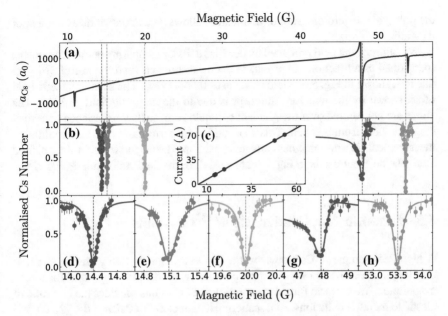

Fig. 5.9 Observation of low-field Cs₂ Feshbach resonances. **a** Dependence of the Cs |3, +3⟩ s-wave scattering length on magnetic field. Theoretical values from [4]. **b** Feshbach spectroscopy of Cs in the 14–60 G region. Normalised Cs number as a function of magnetic field, with 5 Feshbach resonances observed in the 14–60 G region. **c** Calibration of the Bias coil by comparison of the measured resonance positions (in A) to values from literature (in G) [6]. **d–h** Enhanced view of 5 observed Feshbach resonances corresponding to states **d** −2(33)4g(3). **e** −2(33)6g(5). **f** −2(33)4g(4). **g** −2(33)4d(4). **h** x2g(2). Coloured dashed lines indicate the poles of Feshbach resonances extracted from a fit to the spectra

resonance we observe a large reduction in the Cs number, with near 100% contrast in most of the observed features.

We directly measure the location of the Feshbach resonance pole by identifying the atom number minimum. The atom number minimum occurs at the Feshbach pole due to the a^4 scaling of the three-body loss rate. This leads to a maximum in three-body losses where the scattering length diverges. We find the atom number minimum and hence the resonance position B_0 by fitting a Lorentzian profile to the loss feature. The majority of the resonances found are fairly narrow, with a HWHM of $\simeq 100$ mG extracted from the Lorentzian fit. However, this is not a true measure of the resonance width, which is defined as the distance between the pole and the zero-crossing of the scattering length. The broader feature around 48 G (purple), identified as the −2(33)4d(4) state, has a larger width of 0.41(6) G and displays a much larger asymmetry than the other features. The asymmetrical feature arises due to a zero-crossing of the scattering length on the higher-field side of the resonance (see Fig. 5.9a). The reduction in the scattering length near the zero-crossing results in an increase in the Cs number above the number on the low-field side of the resonance where the (background) scattering length is large ($a \simeq 1000\,a_0$). The observation of the pole and the zero-crossing of this feature allows a quantitative measurement of its width.

We calibrate the field produced by our Bias coils by comparing our measurements (performed as a function of current) to the measurements reported in Ref. [6]. The linear fit shown in Fig. 5.9c yields a value of 0.64(1) G/A. The small magnetic field offset shown by the non-zero intercept is due to the magnetic field gradient. The atoms are not centred over the gradient coil field zero, so they experience a residual magnetic field from this coil. This was verified by performing the same calibration measurements in a non-magnetically levitated trap. The magnetic field values plotted along the horizontal axis in Fig. 5.9 are a result of the field calibration in Fig. 5.9c.

5.5   Bose-Einstein Condensate of ^{133}Cs Atoms

A high PSD sample of Cs atoms are trapped in the dimple trap with radial (axial) trap frequencies of 330 (40) Hz. This constitutes an ideal starting point for forced evaporation. We perform forced evaporation at 22 G as this provides the best ratio of elastic to inelastic collisions. At a scattering length of $a_{Cs} = 210\,a_0$ ($B \simeq 22$ G) [25] there exists an Efimov minimum in the three-body recombination rate. Experimentally, the exact magnetic field used during the evaporation is set by optimising the PSD following an evaporation ramp. The change in inelastic loss rate in the region around the Efimov minimum is fairly small and the optimal field value is dependent upon the atomic density which, for example, may require an increase in the elastic collision rate at the expense of a larger inelastic collision rate or vice-versa.

We perform the forced evaporation by lowering the power of both dimple beams over 600 ms, we reduce the power of beam 1 from 80 mW to 20 mW and beam 2 from 200 mW to 50 mW. This first ramp reduces the temperature of the atoms to 300 nK. This is followed by a second forced evaporation ramp lasting 1700 ms, where the power of beam 1 is reduced from 20 mW to 5 mW and beam 2 from 50 mW to 15 mW. These ramps were determined empirically by optimising the efficiency at each stage. A simplified experimental routine showing the steps discussed in this chapter is shown in Fig. 5.10. The whole experimental routine takes 9 s and is mainly limited by the warm up time of the 100 W IPG laser used for the dimple trap. BECs of $2 - 3 \times 10^4$ atoms can be produced in just 4 s but the lack of warm up time for the IPG causes large fluctuations in the number for successive experimental runs.

Following the forced evaporative cooling ramps we observe Bose-Einstein condensation of the Cs atoms. The dimple trap is switched off and the atoms are levitated for a 50 ms time of flight before an absorption image is taken. The BEC transition for Cs is shown in Fig. 5.11. We typically obtain pure condensates of 4–5 $\times 10^4$ atoms, with the onset of degeneracy occurring at $T_c \simeq 60$ nK.

Figure 5.12 shows the evolution of the PSD from the DRSC cooled cloud to BEC. The trap frequencies for the DRSC cooled atoms were estimated from the lattice simulation used earlier to model the Raman lattice. The temperature of the cloud is shown by the red points and the figure clearly shows how the PSD increases

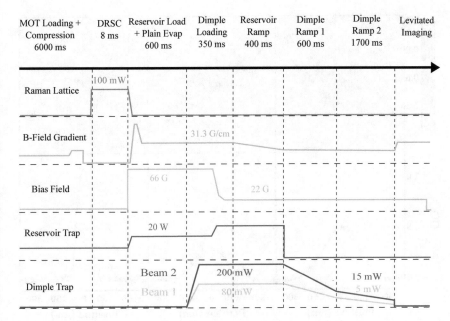

Fig. 5.10 Simplified experimental sequence for the creation of a Cs BEC

dramatically with successive transfers from DRSC to reservoir to dimple even though the temperature of the atoms remains the same. The straight line fit is used to extract the average evaporation efficiency which we find to be $\gamma = 2.5(1)$.

We demonstrate the tunability of the intraspecies interactions of the Cs BEC in Fig. 5.13. We measure the size of the Cs condensate after a levitated time of flight of 70 ms at different magnetic bias fields. In a pure condensate the initial kinetic energy is negligible and the expansion of the cloud is dictated by its mean-field energy [23]. By observing the size of the condensate at a variety of magnetic fields, we explore the tunability of the condensates self-interaction. At each point on the plot a Cs BEC was produced at a bias field of 22 G before the dimple beams were switched off and the bias field jumped to the appropriate value. The BEC was then levitated for 65 ms at the appropriate bias field and subsequently absorption imaged after a further 5 ms of free time of flight.

By tuning the magnetic field in the 15–25 G region we explore three distinct regimes of self-interaction. At negative scattering lengths (fields below 17 G) the condensate collapses due to the strong attractive interactions. This results in a loss of atoms and an increase in the expansion energy of the atoms during time of flight, resulting in a larger width. Around 17 G, where the scattering length crosses zero, the condensate is almost non-interacting. The interaction energy is close to zero which results in only a small expansion of the cloud. For positive scattering lengths (fields above 17 G) the interactions are repulsive resulting in a stable condensate with a

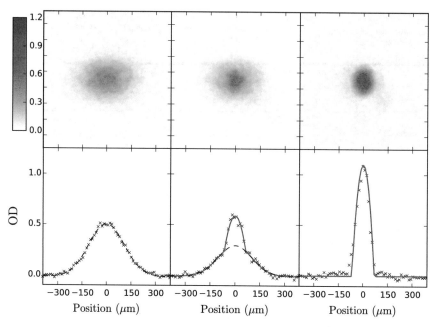

Fig. 5.11 Bose-Einstein condensation of Cs by evaporative cooling. The top panels show absorption images taken after 50 ms of levitated time of flight. The bottom panels show the corresponding horizontal cross-cuts through the images. The laser power of the optical dipole trap is gradually reduced, cooling the atoms from a thermal cloud at $T_{Cs} = 70$ nK (left), across the BEC transition to $T_{Cs} = 50$ nK (middle). Finally, at the end of evaporation a pure condensate (right) is produced containing 5×10^4 atoms

Fig. 5.12 Evolution of Cs PSD (blue) and temperature (red) during the experimental routine. The straight line fit is used to extract an average evaporation efficiency of $\gamma = 2.5(1)$

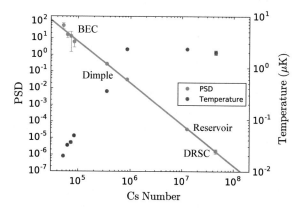

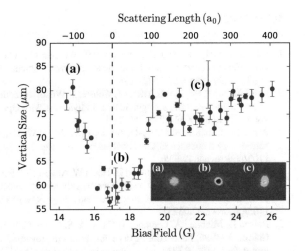

Fig. 5.13 Expansion of a Cs BEC at different magnetic fields. The vertical width of the Cs condensate is plotted as a function of the magnetic bias field applied during the 70 ms levitated time of flight. The inset shows absorption images of the cloud at **a** negative scattering lengths **b** zero scattering length **c** positive scattering lengths

large interaction energy. This large interaction energy results in a large momentum spread of the condensate during time of flight.

The lifetime of the Cs BEC is 3.5(5) s which is about half the lifetime of the Yb BEC. The scattering rate of trap photons at the intensities used are low but it is reasonable to suggest that the shorter lifetime may be due to the higher three-body loss rate in Cs. However, when considering the $\frac{1}{3!}$ reduction in the three-body rate for a degenerate gas [20], the observed lifetime is too short to be explained by three-body loss. The most likely cause of the shortened lifetime is the dimple trapping light provided by the 100 W IPG laser. The high power and large linewidth of these lasers have been observed to cause large loss in Rb due to short range photoassociation [30, 31] and by driving two-photon Raman transitions between hyperfine levels [26]. The losses observed here are evidently not as catastrophic as the Rb case because we are still able to produce a Cs BEC. Yet, these effects could still contribute to a reduced Cs BEC lifetime. In addition, the reduced lifetime may be caused by the increased intensity noise on the trapping beams at the end of the evaporation. Although the dimple beams are stabilised by a feedback servo loop using a two photodiode system, running the IPG laser at 20% of its power requires the AOM to attenuate the beam power by $< \frac{1}{1000}$ which corresponds to very small voltages (with low SNR) applied by the servo to the AOM driver. The fact we can even create a BEC with this level of attenuation suggests that the performance of the intensity servo is more than suitable for the higher powers required to confine Yb.

The above issues could be remedied by using an independent, lower-power, single-mode laser to create the Cs dimple trap. However, the obtainable lifetime of 3.5 seconds was deemed sufficient for the investigation of the interspecies interactions we pursue in the next chapter.

References

1. Anderson MH, Ensher JR, Matthews MR, Wieman CE, Cornell EA (1995) Observation of Bose-Einstein condensation in a dilute atomic vapor. Science 269(5221):198–201. https://doi.org/10.1126/science.269.5221.198
2. Arlt J, Bance P, Hopkins S, Martin J, Webster S, Wilson A, Zetie K, Foot CJ (1998) Suppression of collisional loss from a magnetic trap. J Phys B At, Mol Opt Phys 31(7):L321–L327. https://doi.org/10.1088/0953-4075/31/7/006
3. Arndt M, Dahan MB, Guéry-Odelin D, Reynolds MW, Dalibard J (1997) Observation of a zero-energy resonance in Cs-Cs collisions. Phys Rev Lett 79(4):625–628. https://doi.org/10.1103/physrevlett.79.625
4. Berninger M, Zenesini A, Huang B, Harm W, Nägerl HC, Ferlaino F, Grimm R, Julienne PS, Hutson JM (2013) Feshbach resonances, weakly bound molecular states, and coupled-channel potentials for cesium at high magnetic fields. Phys Rev A 87(3):032,517. https://doi.org/10.1103/physreva.87.032517
5. Boiron D, Michaud A, Lemonde P, Castin Y, Salomon C, Weyers S, Szymaniec K, Cognet L, Clairon A (1996) Laser cooling of cesium atoms in gray optical molasses down to 1.1 μk. Phys Rev A 53:R3734–R3737. https://doi.org/10.1103/PhysRevA.53.R3734
6. Chin C, Vuletić V, Kerman AJ, Chu S, Tiesinga E, Leo PJ, Williams CJ (2004) Precision Feshbach spectroscopy of ultracold Cs_2. Phys Rev A 70(3):032,701. https://doi.org/10.1103/physreva.70.032701
7. Dalibard J, Cohen-Tannoudji C (1989) Laser cooling below the Doppler limit by polarization gradients: simple theoretical models. J Opt Soc Am B 6(11):2023–2045. https://doi.org/10.1364/josab.6.002023
8. Davis KB, Mewes MO, Andrews MR, van Druten NJ, Durfee DS, Kurn DM, Ketterle W (1995) Bose-Einstein condensation in a gas of sodium atoms. Phys Rev Lett 75(22):3969. https://doi.org/10.1103/PhysRevLett.75.3969
9. Fölling S (2003) 3D Raman sideband cooling of rubidium. Master's thesis
10. Gröbner M (2017) A quantum gas apparatus for ultracold mixtures of K and Cs. PhD thesis
11. Grynberg G, Louis B, Verkerk P, Courtois JY, Salomon C (1993) Quantized motion of cold cesium atoms in two- and three-dimensional optical potentials. Phys Rev Lett 70(15):2249–2252. https://doi.org/10.1103/physrevlett.70.2249
12. Guery-Odelin D, Soeding J, Desbiolles P, Dalibard J (1998) Strong evaporative cooling of a trapped cesium gas. Opt Express 2(8):323. https://doi.org/10.1364/oe.2.000323
13. Gustavsson M (2008) A quantum gas with tunable interactions in an optical lattice. PhD thesis
14. Guttridge A, Hopkins SA, Kemp SL, Frye MD, Hutson JM, Cornish SL (2017) Interspecies thermalization in an ultracold mixture of Cs and Yb in an optical trap. Phys Rev A 96(012):704. https://doi.org/10.1103/PhysRevA.96.012704
15. Han DJ, Wolf S, Oliver S, McCormick C, DePue MT, Weiss DS (2000) 3D Raman sideband cooling of cesium atoms at high density. Phys Rev Lett 85(4):724–727. https://doi.org/10.1103/physrevlett.85.724
16. Herbig J (2005) Quantum-degenerate cesium: atoms and molecules. PhD thesis
17. Hopkins SA, Webster S, Arlt J, Bance P, Cornish S, Maragò O, Foot CJ (2000) Measurement of elastic cross section for cold cesium collisions. Phys Rev A 61(3):032,707. https://doi.org/10.1103/physreva.61.032707
18. Hung CL (2011) In situ probing of two-dimensional quantum gases. PhD thesis
19. Hung CL, Zhang X, Gemelke N, Chin C (2008) Accelerating evaporative cooling of atoms into Bose-Einstein condensation in optical traps. Phys Rev A 78(1):011,604. https://doi.org/10.1103/PhysRevA.78.011604
20. Kagan Y, Svistunov BV, Shlyapnikov GV (1985) Effect of Bose condensation on inelastic processes in gases. JETP Lett 42:209
21. Kastberg A, Phillips WD, Rolston SL, Spreeuw RJC, Jessen PS (1995) Adiabatic cooling of cesium to 700 nK in an optical lattice. Phys Rev Lett 74(9):1542–1545. https://doi.org/10.1103/physrevlett.74.1542

22. Kerman AJ, Vuletić V, Chin C, Chu S (2000) Beyond optical molasses: 3D Raman sideband cooling of atomic cesium to high phase-space density. Phys Rev Lett 84:439–442. https://doi.org/10.1103/PhysRevLett.84.439

23. Ketterle W, Durfee D, Stamper-Kurn D (1999) Making, probing and understanding Bose-Einstein condensates, IOS Press, pp 67–176. https://doi.org/10.3254/978-1-61499-225-7-67

24. Kraemer T, Herbig J, Mark M, Weber T, Chin C, Nägerl HC, Grimm R (2004) Optimized production of a cesium Bose-Einstein condensate. Appl Phys B 79(8):1013–1019. https://doi.org/10.1007/s00340-004-1657-5

25. Kraemer T, Mark M, Waldburger P, Danzl JG, Chin C, Engeser B, Lange AD, Pilch K, Jaakkola A, Nägerl HC, Grimm R (2006) Evidence for Efimov quantum states in an ultracold gas of caesium atoms. Nature 440(7082):315–318. https://doi.org/10.1038/nature04626

26. Lauber T, Küber J, Wille O, Birkl G (2011) Optimized Bose-Einstein-condensate production in a dipole trap based on a 1070-nm multifrequency laser: influence of enhanced two-body loss on the evaporation process. Phys Rev A 84(4):043,641. https://doi.org/10.1103/physreva.84.043641

27. Lett PD, Phillips WD, Rolston SL, Tanner CE, Watts RN, Westbrook CI (1989) Optical molasses. J Opt Soc Am B 6(11):2084–2107. https://doi.org/10.1364/josab.6.002084

28. Li Y, Feng G, Xu R, Wang X, Wu J, Chen G, Dai X, Ma J, Xiao L, Jia S (2015) Magnetic levitation for effective loading of cold cesium atoms in a crossed dipole trap. Phys Rev A 91(5):053,604. https://doi.org/10.1103/physreva.91.053604

29. McCarron DJ, Cho HW, Jenkin DL, Köppinger MP, Cornish SL (2011) Dual-species Bose-Einstein condensate of ^{87}Rb and ^{133}Cs. Phys Rev A 84(1):011,603. https://doi.org/10.1103/PhysRevA.84.011603

30. Menegatti CR, Marangoni BS, Bouloufa-Maafa N, Dulieu O, Marcassa LG (2013) Trap loss in a rubidium crossed dipole trap by short-range photoassociation. Phys Rev A 87(5):053,404. https://doi.org/10.1103/physreva.87.053404

31. Passagem HF, Colín-Rodríguez R, da Silva PCV, Bouloufa-Maafa N, Dulieu O, Marcassa LG (2017) Formation of ultracold molecules induced by a high-power single-frequency fiber laser. J Phys B: At, Mol Opt Phys 50(4):045,202. https://doi.org/10.1088/1361-6455/aa5a6e

32. Pinkse PWH, Mosk A, Weidemüller M, Reynolds MW, Hijmans TW, Walraven JTM (1997) Adiabatically changing the phase-space density of a trapped bose gas. Phys Rev Lett 78(6):990–993. https://doi.org/10.1103/physrevlett.78.990

33. Pires RJAA (2014) Efimov resonances in an ultracold mixture with extreme mass imbalance. PhD thesis

34. Söding J, Guéry-Odelin D, Desbiolles P, Ferrari G, Dalibard J (1998) Giant spin relaxation of an ultracold cesium gas. Phys Rev Lett 80(9):1869–1872. https://doi.org/10.1103/physrevlett.80.1869

35. Stamper-Kurn DM, Miesner HJ, Chikkatur AP, Inouye S, Stenger J, Ketterle W (1998) Reversible formation of a Bose-Einstein condensate. Phys Rev Lett 81(11):2194. https://doi.org/10.1103/PhysRevLett.81.2194

36. Steck DA (2010) Cesium D line data. http://steck.us/alkalidata (revision 2.1.4)

37. Thomas AM, Hopkins S, Cornish SL, Foot CJ (2003) Strong evaporative cooling towards Bose-Einstein condensation of a magnetically trapped caesium gas. J Opt B Quantum Semiclassical Opt 5(2):S107–S111. https://doi.org/10.1088/1464-4266/5/2/366

38. Treutlein P, Chung KY, Chu S (2001) High-brightness atom source for atomic fountains. Phys Rev A 63(5):051,401. https://doi.org/10.1103/physreva.63.051401

39. Vuletić V, Chin C, Kerman AJ, Chu S (1998) Degenerate Raman sideband cooling of trapped cesium atoms at very high atomic densities. Phys Rev Lett 81:5768–5771. https://doi.org/10.1103/PhysRevLett.81.5768

40. Weber T (2003) Bose-Einstein condensation of optically trapped cesium. PhD thesis

41. Weber T, Herbig J, Mark M, Nägerl HC, Grimm R (2003a) Bose-Einstein condensation of cesium. Science 299(5604):232–235. https://doi.org/10.1126/science.1079699

42. Weber T, Herbig J, Mark M, Nägerl HC, Grimm R (2003b) Three-body recombination at large scattering lengths in an ultracold atomic gas. Phys Rev Lett 91(123):201. https://doi.org/10.1103/PhysRevLett.91.123201

Chapter 6
Interspecies Thermalisation
in an Ultracold Mixture of Cs and Yb

With suitable techniques developed and implemented for the cooling to degeneracy of Yb and Cs independently, we now begin to explore how to combine the two species in a common optical trapping potential. Confinement of the two species in the same trap enables the study of interspecies interactions and allows the evaluation of potential routes to double degeneracy in the mixture. An important parameter that characterises interspecies interactions is the s-wave scattering length. Knowledge of the scattering length is of tantamount importance for the prediction of interspecies Feshbach resonances and in the appraisal of different approaches for preparing the ultracold mixture.

Similarly to the *intra*-species scattering length, the *inter*-species scattering length has an ideal value in the region of a few tens of bohr radii to a few hundred bohr radii. A scattering length which is too small or too large means that sympathetic cooling of a species will be inefficient due to inelastic losses occurring on similar timescales to thermalisation by elastic collisions. This necessitates the independent cooling of both species and later combination into a combined trapping potential. Additionally, large interspecies scattering lengths may cause a mixture of two BECs to phase separate due to the dominant repulsive interactions. This is problematic for the efficient preparation of molecules as the formation efficiency is related to the overlap of the two species and their phase space densities. This immiscibility problem arises in the RbCs system [20] which necessitates that magnetoassociation be performed on a thermal sample [16, 27]. The creation of a miscible two-species condensate requires that $g_{CsYb}^2 < g_{YbYb} \, g_{CsCs}$, where the interaction coupling constants are [25]

$$g_{ij} = 2\pi\hbar^2 a_{ij} \left(\frac{m_i + m_j}{m_i m_j} \right). \tag{6.1}$$

The magnitude of the scattering length may be calculated from elastic cross sections. Elastic cross sections are measured by observing the thermalisation of two

© Springer Nature Switzerland AG 2019
A. Guttridge, *Photoassociation of Ultracold CsYb Molecules
and Determination of Interspecies Scattering Lengths*, Springer Theses,
https://doi.org/10.1007/978-3-030-21201-8_6

species in thermal contact, an experiment which is usually performed with two species confined in the same trapping potential. This technique is not as accurate as two-photon photoassociation spectroscopy, which we later perform, but still allows the scattering lengths of multiple CsYb isotopologs to be evaluated.

In this chapter we present the preparation of a mixture of Cs and Yb in the dimple trap and first measurements of the scattering properties of ^{133}Cs+^{174}Yb and ^{133}Cs+^{170}Yb. We measure interspecies thermalisation of the mixture in the dimple trap and use a kinetic model to extract effective thermalisation cross sections. Comparison of the extracted cross sections to quantum scattering calculation allows us to evaluate the scattering lengths of all seven CsYb isotopologs. The results of this chapter are published in Ref. [9].

6.1   Preparation of a Mixture of Cs and Yb

The routine for the preparation of both Cs and Yb builds upon the techniques developed for the independent cooling of the two species. However, combined trapping of the two species is more complex than the individual single species approaches due to the contrasting properties of Cs and Yb. Yb has a moderately low polarisability at the trapping wavelength (α_{Yb} (1070 nm) $= 160\,a_0^3$) so that, for the powers used in the thermalisation measurements, Yb atoms are trapped only in the part of the potential where the two dimple beams overlap. Cs, on the other hand, has a much larger polarisability at the trapping wavelength (α_{Cs} (1070 nm) $= 1140\,a_0^3$), creating a trap deep enough that Cs atoms are confined both inside and outside the crossed-beam region of the dimple. Some Cs atoms thus experience a trapping potential dominated by just a single dimple beam. The trapping potential for Cs in the axial direction therefore looks remarkably similar to the combined reservoir and dimple trap used for the preparation of Cs BEC.

A summary of the experimental sequence used for the thermalisation measurements is shown in Fig. 6.1. The two species are sequentially loaded into the dipole trap to avoid unfavourable inelastic losses from overlapping MOTs. We observe a significant drop in the Cs number when the Cs MOT is operated in the presence of the 399 nm Yb Zeeman slower light. This drop in number is likely due to ionisation of Cs atoms in the $6\,^2P_{3/2}$ state. We choose to prepare Yb in the dimple trap first due to the much longer loading time of the MOT and its insensitivity to magnetic fields. We first load the Yb MOT for 10 s, preparing 5×10^8 atoms at $T = 140\,\mu$K [8], before ramping the power and detuning the MOT beams to cool the atoms to $T = 40\,\mu$K. We load 1.8×10^7 atoms into the dimple with a trap depth of $U_{Yb} = 950\,\mu$K. We then evaporatively cool the atoms by exponentially reducing the trap depth to $U_{Yb} = 5\,\mu$K in 7 s, producing a sample of 1×10^6 Yb atoms at a temperature of $T = 550$ nK. This sequence is similar to the preparation of ^{174}Yb BEC in Sect. 4.2.1 but here the evaporation is halted before the BEC phase transition. At this stage the Yb trap frequencies as measured by center-of-mass oscillations are 240 Hz radially and 40 Hz axially.

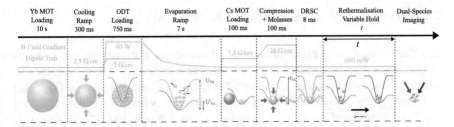

Fig. 6.1 Simplified experimental sequence. The Yb MOT is loaded, then cooled and compressed to facilitate subsequent loading into the dimple trap. The Yb is then evaporated in the dimple by ramping the trap depth until a temperature of $T = 550\,\text{nK}$ is reached. The displaced Cs MOT is loaded before it is compressed, cooled and transferred into a near-detuned optical lattice for DRSC. The DRSC stage loads Cs into the dimple, where it is held with Yb for a variable time t before the trap is switched off and the atoms are destructively imaged after a variable time of flight using dual-species absorption imaging

Once the Yb is prepared in the dipole trap, the Cs MOT is loaded for 0.15 s, at which point the MOT contains 1×10^7 atoms. The Cs MOT is then compressed via ramps to the magnetic field, laser intensity and detuning before it is overlapped with the dimple using shim coils. The Cs atoms are then further cooled by optical molasses before transfer into a near-detuned lattice with $P = 100\,\text{mW}$, where the atoms are then polarized in the $|F = 3, m_F = +3\rangle$ state and cooled to $T = 2\,\mu\text{K}$ with 8 ms of DRSC. During this stage 9×10^4 atoms are transferred into the dimple and the magnetic bias field is set to 22 G, corresponding to the Efimov minimum in the Cs three-body recombination rate [17]. During the transfer the atoms are heated to $T = 5\,\mu\text{K}$. The heating and poor efficiency of the transfer into the dimple are due to the poor mode matching of the DRSC-cooled cloud and the deep dimple trap ($U_{\text{Cs}} = 85\,\mu\text{K}$). The ratio of trap depths $U_{\text{Cs}}/U_{\text{Yb}} = 15.5$ is greater than the ratio of the polarisabilities $\alpha_{\text{Cs}}/\alpha_{\text{Yb}} = 7.1$ due to the effect of gravity on the weak Yb trap. The ratio of the mean trap frequencies between the two species is $\overline{\omega}_{\text{Cs}}/\overline{\omega}_{\text{Yb}} = 3.1$. Given the tightly confining trap for the Cs atoms, we do not use the reservoir trap for the transfer into the dimple as the heating of the sample is large and the overlap with the dimple trap causes large loss of Yb atoms due to sympathetic evaporation. Therefore, to better observe sympathetic cooling we desire a large number imbalance in the favour of Yb.

The thermalisation measurements begin with a mixture of 1×10^6 Yb atoms in their spin-singlet ground state $^1S_0$ and 9×10^4 Cs atoms in their absolute ground state $^2S_{1/2}\,|3, +3\rangle$. For each experimental run the number and temperature are determined by quickly turning off the dimple after a variable hold time and performing resonant absorption imaging of both species after a variable time of flight.

6.2 Interspecies Interactions

In order to simulate sympathetic cooling of two distinct atomic species, a simple kinetic model can be used. Here we will derive a system of 4 coupled equations which describe the evolution of the number (N_{Cs}, N_{Yb}) and temperature (T_{Cs}, T_{Yb}) of the two species. These can then be solved numerically for given initial conditions and be compared with experimental results (see, for example, Fig. 6.2). Differentials with respect to time are denoted with a dot above the symbol, for example $\dot{N}$. Here we draw together different elements from several similar models [2, 13, 14, 19, 22, 24, 28, 29].

We first consider single-species effects, starting with evaporative cooling. If the temperature of the species is not far enough below the trap depth U_i then atoms with sufficient energy may evaporate from the trap. The dimensionless parameter $\eta_i = U_i / k_B T_i$ characterizes the trap depth relative to the temperature. Assuming that $k_B T_i$ is small compared to U_i, atoms with energy greater than the trap depth are produced at a rate of $\Gamma_{ii} \eta_i \exp(-\eta_i)$ [14, 19]. The total effective hard-sphere elastic collision rate Γ_{ii} is conveniently written as $N_i \gamma_{ii}$ where the effective mean collision rate per atom is

$$\gamma_{ii} = \langle n_i \rangle_{sp} \sigma_{ii} \bar{v}_{ii}, \tag{6.2}$$

where σ_{ii} is the elastic scattering cross-section, $\bar{v}_{ii} = \sqrt{16 k_B T_i / \pi m_i}$ is the mean velocity, m_i is the mass of species i and the mean density is given by

$$\langle n_i \rangle_{sp} = \frac{N_i}{8} \left(\frac{m_i \bar{\omega}_i^2}{\pi k_B T_i} \right)^{3/2}, \tag{6.3}$$

where $\bar{\omega}_i = \sqrt[3]{\omega_x \omega_y \omega_z}$ is the mean trap frequency. When an atom evaporates from the trap, it carries away an average energy $\epsilon_{evap} = (\eta_i + \kappa_i) k_B T_i$, where $\kappa_i = (\eta_i - 5)/(\eta_i - 4)$ [24]; this expression for κ_i is appropriate if $k_B T_i \ll U_i$ and U_i is harmonic near the minimum, as is the case in our experiment. The evolution of number and total energy of the ensemble due to evaporation is therefore

$$\dot{N}_{i,evap} = -N_i \gamma_{ii} \eta_i \exp(-\eta_i), \tag{6.4}$$

$$\dot{E}_{i,evap} = (\eta_i + \kappa_i) k_B T_i \dot{N}_{i,evap}. \tag{6.5}$$

The evolution of the temperature can be derived from the relation

$$3 k_B \left(T_i \dot{N}_i + \dot{T}_i N_i \right) = \dot{E}_i, \tag{6.6}$$

to give

$$\dot{T}_{i,evap} = \eta_i \exp(-\eta_i) \gamma_{ii} \left(1 - \frac{\eta_i + \kappa_i}{3} \right) T_i. \tag{6.7}$$

If present, inelastic or reactive two-body collisions could also be included in this model. However, since both Cs and Yb are in their absolute ground state, two-body collisional losses are fully suppressed for our case.

The effect of collisions with background gas is included through the terms

$$\dot{N}_{i,\text{bg}} = -K_{\text{bg}} N_i,$$ (6.8)

$$\dot{E}_{i,\text{bg}} = 3k_B T_i \dot{N}_{i,\text{bg}},$$ (6.9)

where K_{bg} is the background loss rate, which is taken to be the same for both species. There is no corresponding change in temperature as the loss does not preferentially affect either warmer or cooler atoms.

The inclusion of three-body collisions is essential for this system, due to the large three-body loss coefficient in Cs. The loss rate is given by

$$\dot{N}_{i,3} = -K_{i,3} \langle n_i^2 \rangle_{\text{sp}} N_i,$$ (6.10)

where $\langle n_i^2 \rangle_{\text{sp}} = \sqrt{64/27} \langle n_i \rangle_{\text{sp}}^2$ and $K_{i,3}$ is the three-body loss coefficient. Because of the density dependence of three-body collisions, atoms are preferentially lost from the high-density region near the center of the trap. The potential energy in this region is lower than the ensemble average, meaning an average excess energy of $1 k_B T$ remains in the trap for each atom that is lost [29]. In addition to this effect, when three-body recombination produces an atom and a diatom in a state very near threshold, the energy released may be small enough that the atom is not lost, but remains trapped along with a fraction of the energy released [29]. This heating contributes $k_B T_{i,\text{H}}$ per lost atom. The combination of these two effects gives

$$\dot{E}_{i,3} = (2T_i - T_{i,\text{H}}) k_B \dot{N}_{i,3},$$ (6.11)

$$\dot{T}_{i,3} = K_{i,3} \langle n^2 \rangle \frac{T_i + T_{i,\text{H}}}{3}.$$ (6.12)

We now consider interspecies collisions and the thermalisation they cause, as the modelling of these collisions is our primary purpose. The average energy transfer in a hard-sphere collision is [22]

$$\Delta E_{\text{Cs} \rightarrow \text{Yb}} = \xi k_B \Delta T,$$ (6.13)

where $\Delta T = T_{\text{Cs}} - T_{\text{Yb}}$ and

$$\xi = \frac{4 m_{\text{Cs}} m_{\text{Yb}}}{(m_{\text{Cs}} + m_{\text{Yb}})^2},$$ (6.14)

reduces the energy transfer for collisions between atoms of different masses. If the collisions are not classical hard-sphere (or purely s-wave) in nature, then different deflection angles Θ should be weighted by a factor of $1 - \cos \Theta$ [1, 5] and the average energy transferred per collision varies from Eq. 6.13. Such effects are not included explicitly in this simple kinetic treatment, so the resulting cross sections and collision rates should be interpreted as *effective hard-sphere* quantities. We include the effects of deflection angles when we calculate thermalisation cross sections from scattering theory.

In the hard-sphere model, the total energy transferred is just the average energy transferred in a hard-sphere collision multiplied by an effective hard-sphere collision rate Γ_{CsYb}, giving

$$\dot{E}_{Cs,\text{therm}} = -\xi k_B \Gamma_{CsYb} \Delta T, \tag{6.15}$$

$$\dot{E}_{Yb,\text{therm}} = +\xi k_B \Gamma_{CsYb} \Delta T, \tag{6.16}$$

and

$$\dot{T}_{Cs,\text{therm}} = -\frac{\xi \Gamma_{CsYb} \Delta T}{3N_{Cs}}, \tag{6.17}$$

$$\dot{T}_{Yb,\text{therm}} = +\frac{\xi \Gamma_{CsYb} \Delta T}{3N_{Yb}}. \tag{6.18}$$

Since thermalisation collisions do not produce loss, $\dot{N}_{i,\text{therm}} = 0$. We can relate the effective rate to an effective cross section σ_{CsYb} through the relation $\Gamma_{CsYb} = \bar{n}_{CsYb} \sigma_{CsYb} \bar{v}_{CsYb}$. Here,

$$\bar{v}_{CsYb} = \sqrt{\frac{8k_B}{\pi} \left(\frac{T_{Yb}}{m_{Yb}} + \frac{T_{Cs}}{m_{Cs}} \right)}, \tag{6.19}$$

is the mean collision velocity. The spatial overlap $\bar{n}_{CsYb}$ is found by integrating the density distributions of the two species

$$\bar{n}_{CsYb} = \int \left[n_{Cs,\text{single}} (\mathbf{r}) + n_{Cs,\text{cross}} (\mathbf{r}) \right] n_{Yb} (\mathbf{r}) \, d^3r$$

$$= N_{Yb} \frac{m_{Yb}^{3/2} \bar{\omega}_{Yb}^3}{2\pi k_B} \left[\frac{N_{Cs,\text{single}}}{\left(T_{Yb} + \beta_{\text{single}}^{-2} T_{Cs} \right)^{3/2}} \right.$$

$$\left. + \frac{N_{Cs,\text{cross}}}{\left(T_{Yb} + \beta_{\text{cross}}^{-2} T_{Cs} \right)^{3/2}} \right], \tag{6.20}$$

where $\beta_j^2 = m_{Cs} \bar{\omega}_{Cs,j}^2 / m_{Yb} \bar{\omega}_{Yb}^2$, where $j = \{\text{single, cross}\}$ denotes the different cases for Cs atoms trapped in the crossed- and single-beam regions. Equation 6.20 holds true for two clouds centred at the same position. However, due to the large difference in trapping potentials between the two species, Yb experiences a greater gravitational sag than the tightly trapped Cs. For the case of two clouds spatially separated by Δz, the spacial overlap must be reduced by a factor

$$F_z(\Delta z) = \exp\left(-\frac{m_{\text{Yb}}\omega_{\text{Yb},z}^2 \Delta z^2}{2k_B\left(T_{\text{Yb}} + \beta^{-2}T_{\text{Cs}}\right)}\right). \tag{6.21}$$

In our case, where the displacement of the clouds is in the z direction due to gravitational sag and

$$\Delta z = g\left(\frac{1}{\omega_{\text{Cs},z}^2} - \frac{1}{\omega_{\text{Yb},z}^2}\right), \tag{6.22}$$

where g is the acceleration due to gravity.

Combining all these contributions results in coupled equations for the number N_i and temperature T_i of the two species

$$\dot{N}_i = -N_i \gamma_{ii}\eta_i \exp(-\eta_i) - K_{\text{bg}}N_i - K_{i,3}\left\langle n_i^2\right\rangle_{\text{sp}} N_i, \tag{6.23}$$

$$\dot{T}_i = \eta_i \exp(-\eta_i)\gamma_{ii}\left(1 - \frac{\eta_i + \kappa_i}{3}\right)T_i + K_{i,3}\left\langle n_i^2\right\rangle_{\text{sp}}\frac{(T_i + T_{i,\text{H}})}{3}$$
$$\pm \frac{\xi\Gamma_{\text{CsYb}}\Delta T(t)}{3N_i} + \dot{T}_{i,\text{dimple}}, \tag{6.24}$$

where $i = \{\text{Yb, Cs}\}$ and $\langle\ldots\rangle_{\text{sp}}$ represents a spatial average. We choose to neglect the three-body loss coefficient for Yb, $K_{\text{Yb},3}$, because we do not observe any evidence of three-body loss on the experimental timescale in single-species Yb experiments. The Cs three-body loss coefficient is measured to be $K_{\text{Cs},3} = 1^{+1}_{-0.9} \times 10^{-26}\,\text{cm}^6/\text{s}$ at the bias field used in the measurements. In addition to the above terms, $\dot{T}_{i,\text{dimple}}$ is added as an independent heating term to account for any heating from the trapping potential, such as off-resonant photon scattering[1] or additional heating effects due to the broadband nature of the trapping laser [18, 21, 26, 30]. The heating rate for Yb alone is found to be zero within experimental error, so $\dot{T}_{\text{Yb,dimple}}$ is fixed at 1 nK/s, which is the predicted heating rate due to off-resonant photon scattering.

6.3 Measuring the Elastic Cross Section

Figure 6.2 shows the number and temperature evolution of Cs and ^{174}Yb atoms, with and without the other species present. The smaller initial number of Cs atoms is chosen to reduce the density of the Cs such that the effects of three-body recombination play a relatively small role in the thermalisation.[2] Treatment of the number evolution

---

[1]Estimated to be 60 nK/s for Cs using our trap parameters.

[2]We find that a larger initial density of Cs atoms results in a larger final temperature difference between the two species and a greater loss of Cs atoms during the thermalisation.

of the Cs atoms requires careful attention due to the presence of Cs atoms both in the crossed-beam region of the trap and in the wings, where confinement is due to only a single dimple beam. Although the Cs atoms in the crossed- and single-beam regions are in thermal equilibrium, the atoms have different density distributions due to the different potentials experienced. This is an important effect to consider when calculating the spatial overlap of the Cs and Yb atoms.

Atoms in the crossed-beam region are distinguished in absorption images by using a small range of interest (ROI) over the crossed-beam region of the trap and the atom number is extracted from cross-cuts of the image in the axial direction of the trap. We measure the total number of Cs atoms using an ROI over the entire cloud and extract the atom number from cross-cuts of the image in the radial direction of the trap. The number of atoms in the single-beam region is simply the difference between the total number and the number in the crossed-beam region.

We observe an increase in the number of Cs atoms trapped in the crossed-beam region of the trap in the presence of Yb, which we attribute to interspecies collisions aiding the loading of this region. We do not observe any Cs atoms loaded into the crossed-beam portion of the trap in the absence of Yb, so the number is not plotted in this case. For the Cs atoms in the single-beam region, we estimate the axial trapping frequency to be the same as for a single-beam trap, 5 Hz and the radial frequencies to be the same as in the crossed-beam region.

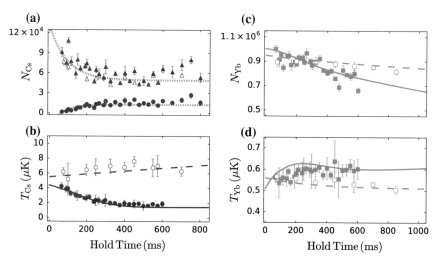

Fig. 6.2 Results of thermalisation experiments. **a** and **b** show the evolution of the Cs number and temperature as a function of hold time t. **c** and **d** show the ^{174}Yb number and temperature as a function of the same hold time. Filled symbols indicate the presence of both Cs and ^{174}Yb in the dimple, whereas open symbols indicate the presence of only one species in the trap. For the Cs number, triangles indicate the number in the single-beam region of the trap and circles the number in the crossed-beam region, while dotted lines show the interpolating functions used to constrain the Cs number in the model. The dashed line shows the result of our kinetic model with only one species trapped and the solid line shows the result for the two-component mixture

We observe a decay of the Yb number throughout the thermalisation. The timescale of this decay is much shorter than single-species $1/e$ background lifetime of 15 s and we attribute the number loss to sympathetic evaporation [23]. The small change in the Yb temperature is explained in part by the evaporation of hotter atoms and also by the large number ratio N_{Yb}/N_{Cs}, which causes the final mean temperature of the sample to be close to the initial Yb temperature. In contrast to Yb, we observe a large change in the temperature of the Cs atoms for short times due to elastic collisions with the Yb atoms. However, for longer times we see the two species reach a steady state at two distinct temperatures. The higher final temperature for Cs results from a Cs heating rate that balances the sympathetic cooling rate.

The coupled equations (6.23) and (6.24) are solved numerically. We perform least-squares fits to the experimental results to obtain optimal values of the parameters σ_{CsYb}, $T_{Cs,H}$ and $\dot{T}_{Cs,dimple}$. The solid lines in Fig. 6.2 show the results of the fitted model, while the dashed lines show the results in the absence of interspecies collisions. Figure 6.2a does not include model results, because our analysis does not include the kinetics of Cs atoms entering and leaving the crossed-beam region. We instead constrain the number of Cs atoms inside and outside this region using interpolating functions (dotted lines in figure) matched to the experimentally measured values.

Since the origin of the heating present on long timescales is unknown, we initially fitted both $T_{Cs,H}$ and $\dot{T}_{Cs,dimple}$. We found that these two parameters are strongly correlated, with a correlation coefficient of 0.99 [10]. We therefore choose to extract the parameter $\dot{T}_{Cs,Heat}$ corresponding to the total heating rate from both recombination heating and heating due to the trap. As shown in Fig. 6.2, the best fit, corresponding to $\sigma_{Cs^{174}Yb} = 5(2) \times 10^{-13}$ cm^2 and $\dot{T}_{Cs,Heat} = 4(1)$ μK/s, describes the dynamics of the system well. The large fractional uncertainty in the value of the elastic cross section is primarily due to the large uncertainty in the spatial overlap. We have investigated the effect of systematic errors in the measured parameters of our model and found that the uncertainty in the trap frequency is dominant and is larger than the statistical error. Inclusion of the correction (Eq. 6.21) is important because the weaker confinement of Yb produces a vertical separation between the two species, reducing the spatial overlap. Initially $F_z(\Delta z) \sim 0.75$. Over the timescale of the measurement the spatial overlap reduces further due to the decreasing width of the Cs cloud as it cools. The final value of $F_z(\Delta z) \sim 0.6$. The effect of gravitational sag on the Cs−Yb overlap is illustrated in Fig. 6.3. It is evident how the overlap will decrease as the Cs cools and the density distribution narrows. The figure also illustrates the contrasting traps for the two species; the very weak Yb trap is strongly effected by gravity whereas the Cs trap provides very strong confinement.

Although the total heating rate extracted from the fit is large, $\dot{T}_{Cs,Heat} = 4(1)$ μK/s, it results from the sum of two heating mechanisms, recombination heating and heating from the optical potential. The value for recombination heating is reasonable because the Cs trap depth of 85 μK is large enough to trap some of the products of the three-body recombination event. For our scattering length, $a_{CsCs} \approx 250\, a_0$, $T_{Cs,H}$ is still within the range of $2\epsilon/9$ and $\epsilon/3$ proposed by the simple model in Ref. [29], where $\epsilon = \hbar^2/m_{Cs}(a_{CsCs} - \bar{a})^2$ with $\bar{a} = 95.5\, a_0$ for Cs. We also cannot rule out any heating

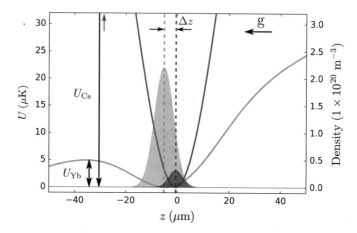

Fig. 6.3 Effect of gravitational sag on the Cs−Yb overlap. Trapping potentials for Cs (red) and Yb (green) in the vertical direction alongside density distributions for the two species after a hold time of 400 ms. Vertical dotted lines denote the centre position of the Cs (red) and Yb (green) clouds

effects due to the broadband nature of the trapping laser [18, 21, 26, 30] which may inflate the value of $\dot{T}_{\text{Cs,Heat}}$ above the simple estimate of 60 nK/s based upon off-resonant scattering of photons. We find that varying the value of the total trap heating rate $\dot{T}_{\text{Cs,Heat}}$ over a large range changes the extracted cross section by less than its error.

For the measurements presented in Fig. 6.2, we deliberately use a low initial density of Cs atoms to avoid three-body recombination collisions dominating the thermalisation. This necessitates use of the weakest possible trap and restricts the number of Cs atoms to 9×10^4. However, due to the large ratio of polarisabilities between Cs and Yb (and the effect of gravity), this results in a very shallow trap for Yb. Preparation of Yb atoms in this shallow trap requires that the intraspecies scattering length be favourable for evaporation. At the time of these measurements the Yb isotopes we were able to study was limited to ^{170}Yb and ^{174}Yb. In Fig. 6.4 we present our thermalisation measurements for ^{170}Yb alongside those for ^{174}Yb. From the fit to the temperature we extract an effective cross section $\sigma_{\text{Cs}^{170}\text{Yb}} = 18(8) \times 10^{-13}$ cm^2 and $\dot{T}_{\text{Cs,Heat}} = 5(2)\,\mu\text{K/s}$. The larger interspecies cross section allows Cs to be cooled to a lower equilibrium temperature. Due to the difference in the natural abundance (31.8% for ^{174}Yb and 3.0% for ^{170}Yb [4]) and the intraspecies scattering lengths ($a_{174} = 105\,a_0$ and $a_{170} = 64\,a_0$ [15]) we obtain a number of ^{170}Yb atoms which is half that of ^{174}Yb, leading to a greater final temperature for ^{170}Yb.

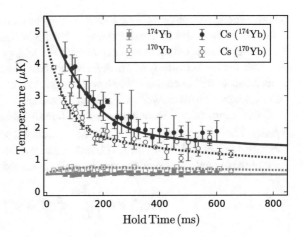

Fig. 6.4 Thermalisation measurements of Cs (red circles) and Yb (green squares) as a function of hold time in the dimple with the other species present. The filled symbols are for ^{174}Yb as coolant and open symbols are for ^{170}Yb as coolant. The solid (dotted) lines shows the best fit of our model for ^{174}Yb (^{170}Yb) isotope as coolant

6.4 Scattering Lengths of CsYb Isotopologs

Except near narrow Feshbach resonances, CsYb collisions can be treated as those of two structureless particles with an interaction potential $V(R)$, which behaves at long range as $-C_6 R^{-6}$. The scattering length for such a system may be related to v_D, the non-integer vibrational quantum number at dissociation, by

$$a = \bar{a}\left[1 - \tan\left(v_D + \tfrac{1}{2}\right)\pi\right],\qquad(6.25)$$

where $\bar{a} = 0.477988\ldots(2\mu_r C_6/\hbar^2)^{1/4}$ is the mean scattering length [7] and μ_r is the reduced mass. For CsYb, with $\sim$70 bound states [3], changes in the Yb isotope alter v_D by less than 1, so that the scattering lengths for all possible isotopologs may be placed on a single curve. Determination of the scattering length for one isotopolog allows predictions for all others.

The connection between scattering lengths and effective cross sections for thermalisation may be made at various different levels of sophistication. At low temperatures, in which atomic scattering takes place purely in the s-wave regime, the cross section between distinguishable particles is

$$\sigma = \frac{4\pi a_{CsYb}^2}{1 + k^2 a_{CsYb}^2},\qquad(6.26)$$

where $k = \sqrt{2E_{th}\mu_r/\hbar^2}$ is the wave vector associated with the collision. Note that the prefactor in the numerator is 4 here as the particles in the Cs+Yb collision are distinguishable. For identical particles, like those considered in the calculation of an intraspecies elastic cross section, the factor would be 8. In the ultracold limit where $ka_{CsYb} \ll 1$; the de Broglie wavelength is much larger than the scattering length. In this regime the cross section is independent of the collision energy $\sigma_{el} = 4\pi a^2$.

Various energy-dependent corrections to σ_{el} may be included, such as effective-range effects or higher partial waves. However, when higher partial waves contribute to the scattering, it is important to replace σ_{el} with the transport cross section $\sigma_\eta^{(1)}$, which accounts for the anisotropy of the differential cross section [1, 5]. p-wave scattering contributes to $\sigma_\eta^{(1)}$ at considerably lower energy than to σ_{el}, because of the presence of interference terms between s waves and p waves.

Calculations of $\sigma_\eta^{(1)}$ were performed by Jeremy Hutson's group at Durham University. The value of $\sigma_\eta^{(1)}$ was calculated explicitly from scattering calculations as described in Ref. [5], using the CsYb interaction potential of Ref. [3]. The resulting energy-dependent cross sections are thermally averaged [1],

$$\sigma_{CsYb}(T) = \frac{1}{2} \int_0^\infty x^2 \sigma_\eta^{(1)}(x) e^{-x} \, dx, \qquad (6.27)$$

where $x = E_{th}/k_B T$ is a reduced collision energy. The thermal average is performed at the temperature $T = \mu_r(T_{Cs}/m_{Cs} + T_{Yb}/m_{Yb})$ that characterizes the relative velocity. Note that Eq. 6.27 contains an extra factor of x because higher-energy collisions transfer more energy for the same deflection angle.

The scattering length of Eq. 6.25 depends only on the fractional part of v_D. Small changes in the integer part of v_D have little effect on the quality of fit. Therefore, the experimental cross sections are fitted by varying the interaction potential of Ref. [3] by the minimum amount needed, retaining the number of bound states. The magnitude of its short-range part is varied by a factor λ to vary the scattering length, while keeping the long-range part $-C_6 R^{-6}$ fixed, with C_6 taken from Ref. [3]. The scattering calculations performed by Matthew Frye and Jeremy Hutson used the MOLSCAT package [12], with the SBE post-processor [11] to evaluate $\sigma_\eta^{(1)}$ from S-matrix elements.

The optimal values of the potential scaling factor λ are found by least-squares fitting to the experimental cross sections. The fit using cross sections from Eq. 6.27 is shown by the solid line in the upper panel of Fig. 6.5. Also shown (dashed line) is a fit using the approximation $\sigma = 4\pi a^2$, with a obtained from scattering calculations using MOLSCAT. The full treatment of Eq. (6.27) gives a better fit than the approximation and it may be seen that there are large deviations between the two approaches for certain reduced masses. These large deviations are shape resonances; a quasibound state with $E > 0$ trapped behind the ℓ-wave (where $\ell = s$, p, d etc.) centrifugal barrier. The p- and d-wave shape resonances that arise in Fig. 6.5 occur when a is close to $2\bar{a}$ and $\bar{a}$ respectively [6].

It is remarkable that even though the temperatures of both species are well below the p-wave barrier height of $\sim 40 \, \mu K$, there is still considerable tunnelling through the barrier, which makes important contributions to the thermalisation cross sections $\sigma_\eta^{(1)}$ because of the interference between s-wave and p-wave scattering.

Typically, measurements of the scattering cross section yield only the magnitude of the scattering length due to the a^2 dependence of the cross section. However, the mass scaling approach allows both the sign and magnitude of the scattering length to be determined from the scaling of the scattering length (cross section) with the

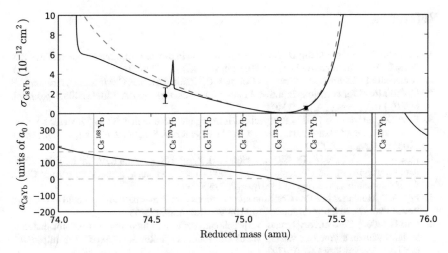

Fig. 6.5 Top panel: thermalisation cross sections as a function of reduced mass, calculated on potentials optimized using the thermally averaged $\sigma_\eta^{(1)}$ given by Eq. 6.27 (black solid line) and the approximation $\sigma = 4\pi a^2$ (dashed line). Points show experimentally measured cross sections and error bars correspond to 1 standard deviation. Bottom panel: Calculated scattering length as a function of reduced mass for the potential optimized using $\sigma_\eta^{(1)}$. Vertical lines correspond to stable isotopes of Yb. Horizontal lines correspond to 0, $\bar{a}$ and $2\bar{a}$

reduced mass of the system. The scattering lengths predicted using the best fitted interaction potential are shown in the lower panel of Fig. 6.5 for all isotopologs of CsYb. The statistical uncertainties in the scattering lengths are quite small, $a = 90(2)$ a_0 for Cs^{170}Yb and $-60(9)$ a_0 for Cs^{174}Yb. The fractional error is smaller for ^{170}Yb than for ^{174}Yb, because Cs^{170}Yb is in a region of reduced mass μ_r where a varies only slowly with μ_r and is mostly determined by $\bar{a}$, which is accurately known. The systematic uncertainties arising from errors in the number of bound states and the kinetic modelling are harder to quantify, but the qualitative features are reliable. The interspecies scattering lengths are moderately positive for Yb isotopes from 168 to 172, close to zero for 173 and moderately negative for 174. The scattering length for Cs^{176}Yb is predicted to be very large, which may produce relatively broad Feshbach resonances [3].

The interspecies scattering lengths obtained above show that both Cs+^{174}Yb and Cs+^{170}Yb BEC mixtures will be miscible at the magnetic field required to minimize the Cs three-body loss rate. Moreover, Fig. 6.5 shows that the interspecies scattering length is predicted to be of moderate magnitude ($<200\,a_0$) for all Yb isotopes except ^{176}Yb. It should thus also be possible to create stable, miscible quantum-degenerate Cs+Yb mixtures for the less abundant ^{168}Yb bosonic isotope and the two fermionic isotopes, ^{171}Yb and ^{173}Yb. Note that the intraspecies scattering length for ^{172}Yb is large and negative [15], precluding the creation of a large condensate.

References

1. Anderlini M, Guéry-Odelin D (2006) Thermalization in mixtures of ultracold gases. Phys Rev A 73(3):032,706. https://doi.org/10.1103/PhysRevA.73.032706
2. Anderlini M, Ciampini D, Cossart D, Courtade E, Cristiani M, Sias C, Morsch O, Arimondo E (2005) Model for collisions in ultracold-atom mixtures. Phys Rev A 72(033):408. https://doi.org/10.1103/PhysRevA.72.033408
3. Brue DA, Hutson JM (2013) Prospects of forming ultracold molecules in $^2\Sigma$ states by magnetoassociation of alkali-metal atoms with Yb. Phys Rev A 87(5):052,709. https://doi.org/10.1103/physreva.87.052709
4. De Laeter JR, Böhlke JK, De Bièvre P, Hidaka H, Peiser HS, Rosman KJR, Taylor PDP (2009) Atomic weights of the elements. Review 2000 (IUPAC technical report). Pure Appl Chem 75:683–800. https://doi.org/10.1351/pac200375060683
5. Frye MD, Hutson JM (2014) Collision cross sections for the thermalization of cold gases. Phys Rev A 89(052):705. https://doi.org/10.1103/PhysRevA.89.052705
6. Gao B (2000) Zero-energy bound or quasibound states and their implications for diatomic systems with an asymptotic van der Waals interaction. Phys Rev A 62(5):050,702. https://doi.org/10.1103/PhysRevA.62.050702
7. Gribakin GF, Flambaum VV (1993) Calculation of the scattering length in atomic collisions using the semiclassical approximation. Phys Rev A 48(1):546–553. https://doi.org/10.1103/PhysRevA.48.546
8. Guttridge A, Hopkins SA, Kemp SL, Boddy D, Freytag R, Jones MPA, Tarbutt MR, Hinds EA, Cornish SL (2016) Direct loading of a large Yb MOT on the $^1S_0 \rightarrow {}^3P_1$ transition. J Phys B: At, Mol Opt Phys 49(14):145,006. https://doi.org/10.1088/0953-4075/49/14/145006
9. Guttridge A, Hopkins SA, Kemp SL, Frye MD, Hutson JM, Cornish SL (2017) Interspecies thermalization in an ultracold mixture of Cs and Yb in an optical trap. Phys Rev A 96(012):704. https://doi.org/10.1103/PhysRevA.96.012704
10. Hughes IG, Hase TPA (2010) Measurements and their uncertainties. Oxford University Press
11. Hutson JM, Green S (1982) SBE computer program. Distributed by Collaborative Computational Project No. 6 of the UK Engineering and Physical Sciences Research Council
12. Hutson JM, Green S (2011) MOLSCAT computer program
13. Ivanov VV, Gupta S (2011) Laser-driven Sisyphus cooling in an optical dipole trap. Phys Rev A 84(6):063,417. https://doi.org/10.1103/PhysRevA.84.063417
14. Ketterle W, Druten NV (1996) Evaporative cooling of trapped atoms. Adv At Mol Opt Phy 37:181–236. https://doi.org/10.1016/S1049-250X(08)60101-9
15. Kitagawa M, Enomoto K, Kasa K, Takahashi Y, Ciuryło R, Naidon P, Julienne PS (2008) Two-color photoassociation spectroscopy of ytterbium atoms and the precise determinations of s-wave scattering lengths. Phys Rev A 77(1):012,719. https://doi.org/10.1103/physreva.77.012719
16. Köppinger MP, McCarron DJ, Jenkin DL, Molony PK, Cho HW, Cornish SL, Le Sueur CR, Blackley CL, Hutson JM (2014) Production of optically trapped ^{87}RbCs Feshbach molecules. Phys Rev A 89(3):033,604. https://doi.org/10.1103/PhysRevA.89.033604
17. Kraemer T, Mark M, Waldburger P, Danzl JG, Chin C, Engeser B, Lange AD, Pilch K, Jaakkola A, Nägerl HC, Grimm R (2006) Evidence for Efimov quantum states in an ultracold gas of caesium atoms. Nature 440(7082):315–318. https://doi.org/10.1038/nature04626
18. Lauber T, Küber J, Wille O, Birkl G (2011) Optimized Bose-Einstein-condensate production in a dipole trap based on a 1070-nm multifrequency laser: influence of enhanced two-body loss on the evaporation process. Phys Rev A 84(4):043,641. https://doi.org/10.1103/physreva.84.043641
19. Luiten OJ, Reynolds MW, Walraven JTM (1996) Kinetic theory of the evaporative cooling of a trapped gas. Phys Rev A 53:381–389. https://doi.org/10.1103/PhysRevA.53.381
20. McCarron DJ, Cho HW, Jenkin DL, Köppinger MP, Cornish SL (2011) Dual-species Bose-Einstein condensate of ^{87}Rb and ^{133}Cs. Phys Rev A 84(1):011,603. https://doi.org/10.1103/PhysRevA.84.011603

21. Menegatti CR, Marangoni BS, Bouloufa-Maafa N, Dulieu O, Marcassa LG (2013) Trap loss in a rubidium crossed dipole trap by short-range photoassociation. Phys Rev A 87(5):053,404. https://doi.org/10.1103/physreva.87.053404
22. Mosk A, Kraft S, Mudrich M, Singer K, Wohlleben W, Grimm R, Weidemüller M (2001) Mixture of ultracold lithium and cesium atoms in an optical dipole trap. Appl Phys B 73(8):791–799. https://doi.org/10.1007/s003400100743
23. Mudrich M, Kraft S, Singer K, Grimm R, Mosk A, Weidemüller M (2002) Sympathetic cooling with two atomic species in an optical trap. Phys Rev Lett 88(253):001. https://doi.org/10.1103/PhysRevLett.88.253001
24. O'Hara KM, Gehm ME, Granade SR, Thomas JE (2001) Scaling laws for evaporative cooling in time-dependent optical traps. Phys Rev A 64(051):403. https://doi.org/10.1103/PhysRevA.64.051403
25. Riboli F, Modugno M (2002) Topology of the ground state of two interacting Bose-Einstein condensates. Phys Rev A 65(6):063,614. https://doi.org/10.1103/PhysRevA.65.063614
26. Sofikitis D, Stern G, Kime L, Dimova E, Fioretti A, Comparat D, Pillet P (2011) Loading a dipole trap from an atomic reservoir. Eur Phys J D 61(2):437–442. https://doi.org/10.1140/epjd/e2010-10261-5
27. Takekoshi T, Debatin M, Rameshan R, Ferlaino F, Grimm R, Nägerl HC, Le Sueur CR, Hutson JM, Julienne PS, Kotochigova S, Tiemann E (2012) Towards the production of ultracold ground-state RbCs molecules: Feshbach resonances, weakly bound states, and the coupled-channel model. Phys Rev A 85(3):032,506. https://doi.org/10.1103/PhysRevA.85.032506
28. Tassy S, Nemitz N, Baumer F, Höhl C, Batär A, Görlitz A (2010) Sympathetic cooling in a mixture of diamagnetic and paramagnetic atoms. J Phys B: At, Mol Opt Phys 43(20):205,309. https://doi.org/10.1088/0953-4075/43/20/205309
29. Weber T, Herbig J, Mark M, Nägerl HC, Grimm R (2003) Three-body recombination at large scattering lengths in an ultracold atomic gas. Phys Rev Lett 91(123):201. https://doi.org/10.1103/PhysRevLett.91.123201
30. Yamashita K, Hanasaki K, Ando A, Takahama M, Kinoshita T (2017) All-optical production of a large Bose-Einstein condensate in a double compressible crossed dipole trap. Phys Rev A 95(1):013,609. https://doi.org/10.1103/physreva.95.013609

Chapter 7
One-Photon Photoassociation

Photoassociation is an important tool in the field of ultracold AMO physics [38]. It extends the fantastic progress of precision spectroscopy in ultracold atoms into the molecular domain. The technique allows precise measurements of molecular potentials useful for characterising atomic interactions and for studies of quantum chemistry [46]. Photoassociation performed in a lattice allows precision spectroscopy of molecular transitions in a Doppler- and recoil-free environment similar to atomic clocks [10, 57]. These molecular clocks are a useful tool for the study of fundamental physics due to their sensitivity to the variation of fundamental constants such as the fine structure constant α and the proton-electron mass ratio [17, 69, 89].

Photoassociation also offers an alternative route to the creation of ground state molecules. Molecules may be produced in a carefully selected electronically excited molecular state with favourable overlap between the ground state wavefunction and the electronically excited wavefunction. This wavefunction overlap, or Franck-Condon factor, can lead to a large number of molecules accumulating in the ground state due to the decay of the excited molecule. This technique has been used to continuously produce ultracold ground state molecules from an atomic sample [2, 3, 24, 42, 70, 88].

There is also burgeoning interest in the all-optical production of ultracold molecules through coherent light-assisted techniques such as Free-Bound STIRAP in an optical lattice [19, 73]. Optical association techniques are the only viable option for the creation of ground state molecules in closed shell systems such as Sr, Yb and Ca. These systems have no Feshbach resonances that can be used to magnetoassociate ground state atoms because of the absence of electronic spin in the atomic ground state. The Free-Bound STIRAP association technique is also of interest to systems where experimentally accessible Feshbach resonances exist but with widths unsuitable for magnetoassociation. Such systems are known to include LiYb and RbYb. However, many systems, including CsYb, require further experimental inves-

© Springer Nature Switzerland AG 2019

A. Guttridge, *Photoassociation of Ultracold CsYb Molecules and Determination of Interspecies Scattering Lengths*, Springer Theses, https://doi.org/10.1007/978-3-030-21201-8_7

tigation before magnetoassociation using predicted Feshbach resonances is verified to be possible.

Photoassociation in a MOT is the most tried and tested method of exploring the vibrational levels of the excited molecular potential. Unfortunately, a dual species MOT of Cs and Yb is difficult to prepare in our current experimental apparatus due to our dual-species Zeeman slower design. While the design of the Zeeman slower allows optimal field profiles for sequential loading of both species, it prohibits loading both simultaneously. The field profiles required for both species are very different and the low capture velocity of the Yb MOT imposes a strict penalty on any deviations from the ideal settings. Dual-species MOT operation is also complicated by balancing the MOT conditions for both species, the contrast is especially stark in the case of alkali-Yb mixtures due to the narrow-linewidth of the Yb MOT, although such problems in alkali-Yb dual species MOTs have been overcome [61, 62, 68]. Further, the 399 nm Yb Zeeman slowing light is seen to ionise the excited state population of the Cs MOT. For these reasons we instead choose to use the more accurate but much more time-consuming method of photoassociation in an optical dipole trap.

In this chapter we present our laser system used for photoassociation of Cs_2 and CsYb molecules. Initially, we investigate the Cs_2 system using our photoassociation setup. Then, photoassociation spectroscopy is used to produce Cs^*Yb molecules and to measure the binding energies of the electronically excited CsYb 2(1/2) state. In addition, we also explore the properties of these long-range molecules. One-photon photoassociation is a critical step towards the two-photon photoassociation measurements that allow the precise determination of the interspecies scattering length. However, the applications of these measurements extend beyond the identification of an intermediate state, as these measurements constitute a precise study of the long-range part of the 2(1/2) state. In the absence of any experimental data on CsYb molecular potentials, the measurements presented in this chapter form a starting point for identification of a two-photon pathway to the CsYb molecular ground state. The Cs^*Yb results presented in this chapter are published in Ref. [35].

7.1   Theory

The creation of CsYb molecules using photoassociation is our first encounter with molecules in the experiment. Therefore, I shall present a brief introduction into the molecular physics required to understand the results later in the chapter.

7.1.1   Molecular Potentials

In the present case we shall consider only diatomic molecules formed from two non-relativistic atoms. By taking into account the electrostatic interactions of the two atoms, one can calculate the potential energy of the atomic pair as a function

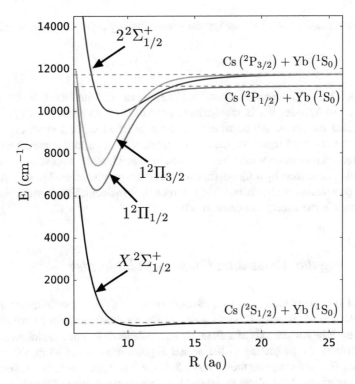

Fig. 7.1 The four energetically lowest Born-Oppenheimer potentials of the CsYb molecule as a function of internuclear distance R. The molecular curves plotted here are adapted from Ref. [58]

of interatomic distance R. An example of these Born-Oppenheimer potentials is shown in Fig. 7.1. These potentials can be broken down into two distinct regions, the short-range region and the long-range region.

In the short-range region the electron clouds are distorted by the presence of the second atom. The interactions here are complex and require the inclusion of many additional atomic configurations to obtain accurate molecular potentials. Models of the short-range potential are therefore limited to approximate methods.

This complexity at short-range is in stark contrast to the long-range behaviour, where the molecular wavefunction may be accurately described by the product of two atomic wavefunctions. The molecules produced through photoassociation may be described as 'Physicist's Molecules' to illustrate their similarity to the case of two free atoms. The properties of these molecules (similar to atomic collisions) are well described by the long-range part of the molecular potential. The molecules produced through photoassociation allow sensitive measurements of the long range part of the molecular potential. The similarity between the photoassociated long-range molecules and two colliding atoms allows photoassociation to be used as a probe of the collisional wavefunction, enhancing our understanding atomic collisions.

The long-range potentials are typically described by asymptotic dispersion coefficients

$$V(r) = -\frac{C_n}{r^n} - \frac{C_m}{r^m} - ..., \tag{7.1}$$

where $V(r)$ is the interaction potential. For two ground state atoms the interaction is limited to Van-der-Waals interactions, described by an attractive C_6/R^6 term. For excited molecular states, where one atom is in an excited state, the leading term is the resonant dipole-dipole C_3/R^3 term, if the two atoms are identical. In heteronuclear molecules where the two atoms are non-identical, the excited molecular potential is described by a C_6 coefficient like in the ground state. The difference in excited potentials between homo- and heteronuclear molecules leads to considerable differences in the photoassociation spectra.

7.1.2 Angular Momentum Coupling in Molecules

It is typical in many molecules for both electronic orbital and electronic spin angular momentum to be present. The procession of the electronic angular momentum, L, around the internuclear axis of a diatomic molecule leads to a new quantum number, Λ, describing the projection of the orbital angular momentum along the internuclear axis. The spin angular momentum, S, is the total spin quantum number of the constituent atoms. In bi-alkali systems the different orientations of the atomic spins can produce singlet, $S = 0$, and triplet, $S = 1$, molecular potentials in the ground state. The presence of these two electronic molecular potentials is important for the existence of broad Feshbach resonances, due to the strong coupling between the two potentials [18]. The projection of the spin angular momentum along the internuclear axis is described by the quantum number Σ. The component of the total angular momentum along the internuclear axis is Ω, where $\Omega = \Lambda + \Sigma$. Of course, in addition to the couplings along the internuclear axis, the molecule may also rotate. This rotation, by definition occurs perpendicular to the internuclear axis and introduces a rotational angular momentum, ℓ.

The presence of the many different angular momenta gives rise to questions concerning the coupling between the various angular momenta. The couplings between these different angular momenta were first outlined by Hund in 1926. Here, we briefly review some of the coupling cases pertinent to this work and point the interested reader to other texts [12, 36] where the various coupling cases are presented in much finer detail.

The Born-Oppenheimer curve of the CsYb ground state in Fig. 7.1 is labelled as $X\,^1\Sigma_{1/2}^+$. This notation corresponds to

$$^{2S+1}\Lambda_\Omega^{+/-},$$

where $+/-$ is the parity of the electronic wavefunction upon reflection at an arbitrary plane containing the internuclear axis and is only used when describing states with no angular momentum component along the internuclear axis ($\Lambda = 0$, Σ states). This notation is used to describe Hund's case (a) and Hund's case (b), both occur for a tightly bound molecule in the short-range part of the potential, $R < 20\,a_0$. At short range the splittings between the Born-Oppenheimer potentials are large compared to relativistic spin-orbit interactions and rotational energies. Hund's case (a) is the relevant coupling case for Cs_2 and excited states of CsYb at short range as the spin-orbit interaction is stronger than the rotational interaction. Molecules in which the opposite is true, the rotational interaction is stronger than the spin-orbit interaction, are described by Hund's case (b). This is the relevant case for the $X\,^1\Sigma_{1/2}^+$ ground state of CsYb.

For molecules produced at long range, for example through photoassociation, the most common coupling case is Hund's case (c), where the spin-orbit interaction is dominant over other the Born-Oppenheimer and rotational interactions. This is usually the case for heavy alkali atoms such as Cs. The states are labelled by

$$\Omega_{u/g}^{+/-}$$

where u/g denotes the parity of the electronic wavefunction upon reflection around the molecular origin and applies only to homonuclear molecules. In Hund's case (c) Ω and J are good quantum numbers, J is formed from the coupling of Ω to the molecular rotation ℓ. The molecular potentials below the Cs D_1 and D_2 lines are presented in Fig. 7.2 as an example of Hund's case (c). An interesting feature of Cs_2 is the 0_g^- (and also 1_u) 'long-range potential' which disassociates at the D_2 line. This potential features an external long-range well in which a molecule excited to this potential oscillates between long-range and intermediate distances where the overlap with the electronic ground state potential is larger. This state and the long-range 1_u state have been studied extensively in an effort to produce ultracold ground state molecules through photoassociation [21, 28, 51, 52, 54, 55].

The long-range CsYb molecules produced through photoassociation are described by the rare Hund's case (e), a case for which to the author's knowledge there have only been a handful of observations [7, 14, 15, 62]. In fact this case was not originally identified by Hund but was introduced later by Mulliken [60]. In Hund's/Mulliken's case (e) neither L nor S couple to the internuclear axis. The coupling is dominated by spin orbit coupling between L and S forming J_a which then couples to the molecular rotation. J_a and ℓ are good quantum numbers and the rotational progression is described by $E_{rot} = B_{rot}\ell(\ell + 1)$.

It is important to note that the cases outlined above are simple, limiting cases useful for approximating the molecular state. It is evident that as two atoms approach each other from a large distance the presence of atom B will have a gradually increasing effect on atom A. Therefore, the Born-Oppenheimer, spin-orbit and rotational interactions will all gradually change as a function of internuclear separation and

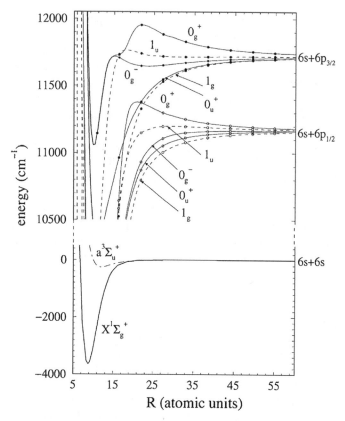

Fig. 7.2 Cs_2 potentials near the $^2S +^2 P_j$ asymptote. At long range the fine structure splitting is large compared to the Born-Oppenheimer interaction. Therefore, the multiple excited molecular potentials are labelled using Hund's case (c). The ground state triplet ($a\,^3\Sigma_u^+$) and singlet ($X\,^1\Sigma_g^+$) potentials are labelled using Hund's case (a). Reprinted from Advances in Atomic, Molecular and Optical Physics, Volume 47, Françoise Masnou-Seeuws and Pierre Pillet, Formation of ultracold molecules ($T \le 200\,\mu K$) via photoassociation in a gas of laser-cooled atoms, 53–127, Copyright (2001) [56], with permission from Elsevier

the actual angular momentum coupling corresponds to an intermediate regime. In the intermediate regime, the coupling is described only by whichever basis is most convenient for defining a Hamiltonian and calculating its matrix elements and eigenvalues.

7.1.3   Vibration

The attractive electronic potentials, $V_i(R)$, illustrated in Fig. 7.1 support a number of bound states. The wavefunctions, $\Psi_i(R)$, and binding energies of these states, E, may be found by solving the Schrödinger equation

$$\left(\frac{\hbar^2}{2\mu_r}\frac{d^2}{dR^2} + V_i(R)\right)\Psi_i(R) = E\Psi_i(R), \tag{7.2}$$

where μ_r is the reduced mass. The model wavefunctions shown in Fig. 7.3 are found by solving Eq. 7.2 using the Leonard-Jones form of the $1\,^1\Pi_{1/2}$ potential given in Ref. [58]. The solutions correspond to various vibrational levels of the molecular potential and are labelled by the vibrational quantum number v, numbered up from zero at the bottom of the potential. The vibrational number corresponds to the number of nodes in the wavefunction, analogous to the case of a quantum harmonic oscillator. Indeed, a diatomic molecule is typically used as a textbook example of a quantum harmonic oscillator and the wavefunctions of the lower vibrational levels are reminiscent of its solutions. However, as the vibrational number increases and the potential becomes more asymmetric, the molecular vibration becomes anharmonic and the wavefunctions deviate from simple harmonic oscillator solutions, becoming more asymmetric. This asymmetry is especially prominent at the outer turning points of the potential, the distance R_t. The molecules in higher vibrational levels spend the majority of their time at large internuclear separations around R_t. Therefore, the properties of molecules in these states are dictated by the long range part of the molecular potential. Photoassociation is most likely to occur at the outer turning point of the potential where the vibrational wavefunction is peaked. As a result, molecules produced through photoassociation are preferentially produced in the higher vibrational levels of the potential, making it an effective technique for precisely studying the long range part of the molecular potential.

The long-range coefficients may be extracted from photoassociation (PA) spectra using near-dissociation expansion formulas. The simplest and most widely used of these expansions is the Le Roy-Bernstein (LRB) formula [48] which links the energy E_v of the vibrational state v to the asymptotic form $(D - C_n/R^n)$ of the potential

$$E_v \simeq D - \left(\frac{v_D - v}{B_n}\right)^{2n/(n-2)},$$

$$B_n = \sqrt{\frac{2\mu_r}{\pi}}\frac{C_n^{1/n}}{\hbar(n-2)}\frac{\Gamma(1/2 + 1/n)}{\Gamma(1 + 1/n)}, \tag{7.3}$$

where v_D is the non-integer vibrational quantum number at dissociation, D is the threshold energy, Γ is the gamma function and μ_r is the reduced mass. At long range, the CsYb electronically excited potentials are dominated by the van der Waals $n = 6$ term. In practice, it is more convenient to express $v_D - v$ in terms of n' and v_{frac}, the fractional part of v_D. This is because in one-photon photoassociation we observe lines near threshold, so it is simpler to label vibrational levels by their number below threshold, starting from $n' = -1$. Explicitly, $n' = v - v_{max} - 1$, where v_{max} is the vibrational quantum number of the least-bound state. As the levels we observe are all close to threshold, n' is relatively easy to determine, but we cannot label the states by v as v_{max} is initially unknown. For a single isotope, v_{max} does not affect the predicted level positions.

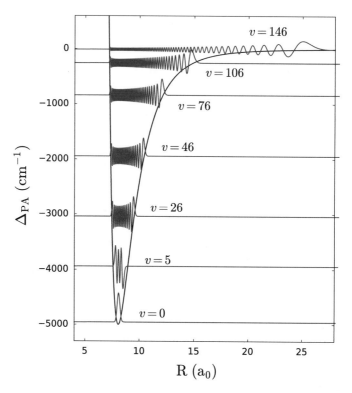

Fig. 7.3 Vibrational wavefunctions of a molecular potential. The vibrational wavefunctions are found by solving Eq. 7.2 for a Lennard Jones potential with similar depth and dispersion coefficients to the CsYb $1\,^1\Pi_{1/2}$ potential. Binding energies of some levels have been offset for illustrative purposes

7.1.4 Photoassociation Transition Strengths

In photoassociation, the transition strength is dictated by the molecular Rabi frequency, $\Omega_{mol} = \langle g | \mathbf{d}_{mol} \cdot \mathbf{E} | e \rangle / \hbar$. This describes the coupling between the initial scattering state $|g\rangle$ and the vibrational level of the excited state $|e\rangle$ induced by the PA light. For vibrational levels with an outer turning point R_t at long range, the electron clouds of the molecule's constituent atoms are well separated and we may describe the molecular Rabi frequency as

$$\Omega_{mol} = \sqrt{f_{rot}}\sqrt{f_{FC}}\,\Omega_{atom}. \tag{7.4}$$

where Ω_{atom} is the Rabi frequency of the atomic transition and f_{rot} is the rotational factor which accounts for rotational couplings and depends upon the molecular states addressed and the polarisation of the PA light. The strength of vibrational transitions is determined by the f_{FC} term, which is known as the Franck-Condon factor

$$f_{FC} = \left| \int_0^\infty \Psi_g(R)\Psi_e(R)dR \right|^2, \qquad (7.5)$$

which describes the overlap between the initial and final state wavefunctions $\Psi_g(R)$ and $\Psi_e(R)$. For excited and ground state potentials with different asymptotic forms (i.e for homonuclear molecules with R^{-3} and R^{-6}), an analytical solution for Eq. 7.5 may be found using the reflection approximation [8, 9, 39]. The premise of this approach is that the rapid oscillations of the integrand in Eq. 7.5 keeps contributions to the integral near zero except at the point of stationary phase. This is located about the Condon point, R_c, which is the internuclear distance where the difference between the excited and ground state potentials equals the photon energy, $V_e(R_c) - V_g(R_c) = \hbar\omega$. For an excited molecular potential with a R^{-3} asymptotic form, as in a homonuclear molecule, the potential extends to a much longer range than the ground state potential which has R^{-6} form. Therefore, for calculating the Condon point of a photoassociation transition, it is a very good approximation to neglect the ground state potential. This leads to the Condon point being located at the outer turning point of the excited potential, $R_c = R_t$.

The formula for the free-bound Franck-Condon factor using the reflection approximation is [9]

$$f_{FC} = \frac{\partial E_v}{\partial v} \frac{1}{D_C} |\Psi_g(R_c)|^2, \qquad (7.6)$$

where

$$D_C = \left| \frac{d}{dR} \left[V_e(R) - V_g(R) \right] \right|_{R=R_c}$$

is the slope of the difference potential evaluated at the Condon point and $\frac{\partial E_v}{\partial v}$ is the vibrational spacing between adjacent levels in the excited state. The reflection approximation suggests that the strength of a transition is simply dictated by the amplitude of the ground state scattering wavefunction at the turning point of the excited molecular potential. This observation allows the atomic scattering length to be calculated from PA spectra [1, 32, 59, 78, 80]. Exciting transitions to different vibrational levels of the excited state creates molecules at different internuclear separations, as the turning point R_t of the potential changes with the binding energy of the vibrational level. Therefore, using the transition strengths of PA spectra the amplitude of the ground state scattering wavefunction can be mapped out as a function of internuclear distance R. Modelling the scattering wavefunction mapped out using this method allows the atomic scattering length to be extracted.

7.2 Experimental Setup

The laser setup we use to perform PA spectroscopy in the dimple trap is displayed in Fig. 7.4. In PA spectroscopy the observable vibrational levels of the electronically excited molecular potential are typically detuned by 10's to 100's of GHz from atomic

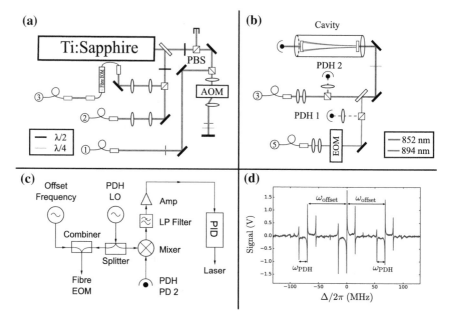

Fig. 7.4 One-photon photoassociation laser setup. **a** Ti:S laser setup on the PA optical table. Fibre 1 carries light to the main experiment and Fibre 2 to the wavemeter. **b** Stabilisation of photoassociation light to a high finesse cavity on the laser table. The dashed lines show the reflected light from the cavity that is picked off using a PBS. Fibre 5 outputs 852 nm light from the Cs repump laser. **c** Electronics for PA locking. The electronics are composed of the standard components used in a PDH locking scheme, with the addition of an 'Offset Frequency', ω_{offset}, which is combined with the usual 'PDH Frequency', ω_{PDH}, used for modulation of the EOM. **d** PDH locking signal using the 'electronic sideband' technique. Zero detuning is set at the frequency of the cavity peak. The additional sidebands at $\Delta = \pm\omega_{\text{offset}}$ are the sidebands we stabilise to the cavity, allowing the tuning of the PA frequency via ω_{offset}. These 'offset' sidebands also show the typical PDH lineshape, with the 'sub-sidebands' at $\Delta = \pm\omega_{\text{offset}} \pm \omega_{\text{PDH}}$

resonance, therefore, we require a laser setup in which the PA laser may be scanned over a large frequency range. For the PA laser we use a highly tunable Ti:Sapphire laser (M Squared SolsTiS).

For photoassociation in the dimple trap we require the PA laser's frequency to be stable and reproducible throughout many experimental runs. Therefore, the PA laser is stabilised to the transmission peak of a high-finesse optical cavity using the Pound-Drever-Hall technique (PDH) [6, 26]. The high-finesse optical cavity used in this work is home-built, with the construction and characterisation detailed in the thesis of Köppinger [44]. The free spectral range (FSR) of the cavity is measured to be $\nu_{\text{FSR}} = 748.852(5)$ MHz. This was measured by scanning the modulation frequency of the fibre-coupled EOM while monitoring the cavity transmission (for more information on this technique see Ref. [34]). Although the cavity mirrors are under vacuum and separated using a Zerodur spacer, the cavity still exhibits a drift rate of > 10 MHz/hr. This is most likely due to the piezo stack which is mounted on one of the cavity mirrors for control of the cavity length. For long experimental runs this drift is larger than the natural linewidth of the transitions we hope to measure, therefore, we stabilise the cavity length using the PDH technique. We use 852 nm light from the zeroth order of the Cs ZS repump AOM and mode match this into the cavity. The PDH locking scheme generates an error signal that is fed into a PID controller. The PID controller provides feedback to the piezo stack which controls the cavity length. The 852 nm light used in the PDH scheme is stabilised to a Cs atomic transition using FM spectroscopy as described in Sect. 3.3.1, therefore, the cavity inherits the long-term stability of the Cs repump lock.

The optical layout of the setup on the PA optical table and cavity table is shown in Fig. 7.4a, b respectively. On the PA laser table a pickoff extracts a small fraction of light for frequency stabilisation and absolute frequency calibration of the PA light. PA light is sent to the optical cavity via a broadband fibre-coupled EOM (EOSPACE PM-0S5-10-PFA-PFA-895) which modulates the light with frequency sidebands. We utilise the 'electronic sideband' technique [34, 49, 79] to allow continuous tunability of the PA laser frequency; by stabilising one of the sidebands to a cavity transmission peak, the frequency of the carrier may be tuned over the 748.852(5) MHz FSR of the cavity by changing the modulation frequency applied to the EOM. The electronics used in this technique is shown in Fig. 7.4c, where PDH LO is the local oscillator used to provide the PDH sidebands for the locking scheme (Fig. 7.4d). We note that for $\omega_{\text{offset}} < \omega_{\text{PDH}}$ or $\omega_{\text{offset}} = \omega_{\text{FSR}}$ the lock is very unstable. If we require PA light at these frequencies we must shift the cavity length by stabilising the Cs repump light to another cavity mode, this shifts the required ω_{offset} by $\nu_{\text{FSR}} \left(1 - \frac{\lambda_{\text{Repump}}}{\lambda_{\text{PA}}} \right) \approx$ 35 MHz for each cavity mode. Precise frequency calibration with respect to the Cs D_1 transition is then achieved by counting cavity fringes from the D_1 transition and including the RF modulation offsets of the carrier. In practice a commercial wavemeter (Bristol 671A) is used to identify the specific cavity fringe used to stabilise the PA laser frequency.

The majority of the Ti:Sapphire laser output is doubled-passed through an AOM to allow control of the intensity of the light and to allow the frequency of the light to be spectrally broadened by dithering the AOM frequency. As the separation between vibrational levels is typically >10 GHz, spectrally broadening the light reduces the number of steps required to cover this range, making scanning less time-consuming. After the AOM the light passes through a beam shutter and into an optical fibre which delivers light onto the experimental table. The output of the fibre passes through a quarter-wave plate to allow for control of the polarisation. In the linear configuration the light is vertically polarised, this allows π transitions to be driven when the magnetic bias field is oriented in the vertical direction, this is the standard configuration in our experiment. After the waveplate the PA light is counterpropagated against the dimple beam 1 using a dichroic mirror and is focused onto the atoms with a waist of 150 μm.

7.3   Photoassociation of Cs_2

As observation of heteronuclear photoassociation is a challenging endeavour, we begin our photoassociation measurements with the somewhat simpler homonuclear case. We perform photoassociation spectroscopy on Cs_2, a molecule that has been extensively studied in the past. The groups of Pierre Pillet in Paris [21, 25, 27, 52, 53, 56, 81, 82, 84, 85] and William Stwalley in Connecticut [64–66] have performed numerous experiments on Cs_2 using the photoassociation technique. Also, at the group of Hanns-Christoph Nägerl in Innsbruck, Cs_2 was among the first molecules produced in the rovibrational ground state [22, 23]. The comprehensive list of binding energies for levels below the Cs D_1 line presented in Ref. [64] was particularly useful when searching for the first features.

Below we first describe preliminary experiments in a MOT before moving on to measurements in the dipole trap, which is required for the interspecies study presented later.

7.3.1   Measurements Using a Cs MOT

The majority of photoassociation studies take place in a MOT, PA transitions in a MOT may be identified using a variety of techniques, the two most popular being ion detection and trap-loss spectroscopy. In ion detection, resonances are identified by the detection of molecular ions produced by ionising the photoassociated molecule using an ionisation laser [50]. This technique is very sensitive due to the near-zero background and allows for high scan rates. However, ion detection requires the installation of an ion-detector, such as an MCP, in the experimental setup which would require significant changes to our apparatus. We opt to use the much simpler trap-loss spectroscopy method for this investigation. In trap-loss spectroscopy the atom number of the MOT is continuously measured using the atomic fluorescence. On a PA resonance an excited molecule is produced, subsequently the molecule spontaneously decays into a ground state molecule or into two free atoms. In both cases there is a dip in the atomic fluorescence as the molecule goes 'dark' to the MOT beams or when the two free atoms are lost from the trap due to the release of the molecular binding energy (typically >1 K). Continuous monitoring of the atom number enables continuous scanning of the PA laser frequency across a large range, which allows an extensive survey of vibrational levels.

The strength of the PA transition is dictated by the Franck-Condon factor, which, as we discussed earlier, is governed by the amplitude of the scattering wavefunction at the Condon point. For homonuclear molecules like Cs_2 the outer turning point of the potential is effectively equal to the Condon point, $R_t = R_c$. High-lying vibrational levels whose Condon point lie at large interatomic distances have much stronger transitions than more deeply bound vibrational levels, which occur further detuned from resonance. This would suggest that in the first instance we should search with a small detuning from resonance, as the transitions are stronger. However, lines close to resonance are difficult to observe due to the high spectral density of the lines and the large depletion of the MOT due to the small detuning from atomic resonance. Deeply-bound levels far detuned from the atomic transition avoid this problem but suffer from weaker transition strengths. The observation of weak transitions far detuned form resonance may be overcome by using more sensitive fluorescence detection methods [55, 87] or by instead detecting ionised molecules using a MCP. However, for the preliminary investigation we address weakly-bound states near dissociation, yet sufficiently detuned such that the off-resonant scattering of PA light is non-problematic.

The detection of PA is complicated by the variety of other loss mechanisms present in a MOT, many of which can overshadow the PA loss. In the investigation of Cs$_2$ PA, we illuminate the Cs MOT with PA light focused to a waist of 150 μm at the MOT centre. The MOT fluorescence is detected using a high gain fluorescence detector (same design as used for the Yb spectroscopy in Sect. 3.3.2) mounted onto an Yb viewport. The temporal dependence of the atom number in a MOT illuminated by PA light is described by the rate equation

$$\dot{N} = L - K_{bg}N - (\beta + \beta_{PA}) \int_V n^2(\mathbf{r}) d^3\mathbf{r}, \qquad (7.7)$$

where N is the number of atoms in the MOT, n is the density of atoms in the MOT, L is the loading rate, K_{bg} is the loss rate due to background collisions, β is the loss rate due to binary (non-PA) collisions among the trapped atoms and β_{PA} is the loss rate from photoassociation of the trapped atoms. The PA signal can be maximised by increasing the PA loss rate over the other loss processes. This can be accomplished by increasing the intensity of the PA light at the atoms. Therefore, we retroreflect the PA light back through the MOT to increase the PA intensity, we obtain a peak intensity of 250 W/cm^2. In addition, the MOT is compressed by increasing the axial gradient to 25 G/cm, which increases the density and increases the number of atoms which interact with the most intense part of the PA beam. The increased gradient also serves to reduce the recapture range of the MOT, an essential step to stop the recapture of dissociating molecules. Also, the atomic beam shutter is partially closed, reducing the number of Cs atoms travelling down the Zeeman slower towards the MOT. This results in a smaller loading rate but also reduces the background collisional loss rate. These changes allow operation in a regime where two-body loss is dominant. However, two-body loss due to PA is not the only two-body process present. Light-assisted collisions cause a large amount of two-body loss in a MOT, these may be mitigated by reducing the intensity of the MOT cooling light, which we reduce to $I_{tot} = 10$ mW/cm^2. This reduces loss through light-assisted collisions but it remains an important loss process in the system, especially when the PA laser is close to atomic resonance.

Figure 7.5 shows the photoassociation of Cs$_2$ molecules in a Cs MOT. The experimental setup for this preliminary investigation is much simpler than the one described in the preceding section. A Distributed-Bragg-Reflector (DBR) laser is focused onto

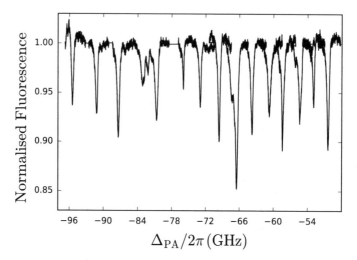

Fig. 7.5 Cs_2 photoassociation in a Cs MOT. The MOT fluorescence is plotted as a function of detuning from the Cs D_1 transition

a Cs MOT and its frequency is scanned in 3 GHz steps with the frequency calibrated using a wavemeter. Each scan was repeated three times and the obtained fluorescence spectrum averaged. It is essential to scan the photoassociation laser frequency slowly (2 GHz/min in this case) to observe these features. The slow scan allows the photoassociation beam to be resonant for approximately the response time of the MOT, which is around 2 s in this case. When the photoassociation beam is resonant with a transition to a bound state, a rapid depletion of the MOT occurs. As the detuning of the photoassociation beam increases, the loss rate decreases and the MOT number recovers. The width and shape of the features are inherently linked to the scan rate of the laser frequency because the rate of increase in fluorescence away from resonance is limited by the MOT properties. The long timescale for depletion and reloading of the MOT also contributes to the asymmetry of the loss features.

The measured line positions are in moderate agreement with the measurements of Pichler et al. [64]. The discrepancies between our data and that of [64] may be due to this lag in the response of the MOT or due to thermal noise on the DBR laser diode over the long scan times.

7.3.2 Optically Trapped Cs

The extension of our current experimental setup to the creation of a dual species MOT of Cs and Yb is far from trivial, so here we shall also investigate the possibility of observing photoassociation of an optically trapped sample. Photoassociation in a dipole trap offers many distinct advantages over that of photoassociation in a MOT.

Firstly, in a dipole trap the atoms have a much higher density and lower background loss rates which yield much higher SNR features. The low background loss rates allows longer interrogation times supporting the investigation of deeper bound levels using the trap loss technique [45]. Secondly, the observed lines have a greater spectral resolution than in a MOT due to the lower power required to observe the features and the absence of thermal broadening (in the alkali case) for the low temperatures achievable following evaporation. The linewidths of PA transitions approach the natural linewidth, facilitating measurements of the binding energy with increased precision. Finally, the well defined initial state of the atoms in a dipole trap allows easier identification of the hyperfine and rotational structure of the PA features.

The main drawback of photoassociation in a dipole trap is the long cycle time of the experiment. In a MOT the constant readout of the atom number through atomic fluorescence allows the PA laser to be continuously scanned. However, in a dipole trap we destructively measure the atom number using absorption imaging after illumination with PA light, meaning the same atomic sample cannot be recycled and must be reloaded each time. The frequency of the PA laser must be painstakingly stepped for each experimental iteration which drastically reduces the frequency range explored using this technique. In addition, systematic effects such as light shifts from the trapping laser must be accounted for when measuring binding energies.

We prepare a sample of Cs atoms using the same initial steps of the routine used for the preparation of the Cs BEC in Chap. 5. However, in contrast to the Cs BEC routine we do not evaporatively cool the Cs atoms in the dimple trap. Once loaded into the dimple and after a short hold to allow the atoms to thermalise, we perform photoassociation spectroscopy on a thermal sample of 2×10^5 Cs atoms at $T = 2\,\mu K$. PA spectroscopy is performed by illuminating the Cs sample with linearly polarised photoassociation light for 300 ms. We keep the bias field at 23.9(5) G to suppress loss due to three-body collisions. After the photoassociation pulse we release the Cs atoms from the dimple and image them using standard absorption imaging.

In the initial scans searching for a Cs_2 feature we noticed that the lifetime of the Cs atoms during interrogation with off-resonant PA light was much shorter than expected for loss caused by heating out of the trap from the off-resonant scattering of photons. To investigate this loss we observe the Cs atom loss as a function of PA pulse time for off-resonant light. We see a drastic increase in the loss rate for detunings closer to the Cs D_1 line. For all detunings the loss rate is much higher than we would expect from a constant heating rate due to off-resonant scattering of photons. Therefore, we attribute the loss to be light-induced trap loss due to excitation of the colliding Cs atoms to a 'quasi-molecular' state below the $6P_{1/2}$ potential [40]. It is important to note that this is not photoassociation to a bound vibrational state. We will briefly summarise the process here but for more information on these collisional processes see Ref. [86]. After excitation by the PA light detuned from an atomic transition, the decay of the quasi-molecular state proceeds through the Radiative Escape mechanism described in the literature, where the decay of the excited atoms produce energetic ground state atoms. The kinetic energy (KE) of the ground state atoms is dependent upon the KE gained by the quasi-molecule as it was accelerated along the molecular potential towards smaller R. If this energy is larger than the trap depth U then the atom is lost. For KE lower than U it remains trapped but its KE is deposited in the sample, raising its temperature. The states populated by the decay of the quasi-molecule are dictated by its branching ratios. In the case considered here, the detuning of the light is fairly large compared to the detunings investigated in the majority of the literature, which is mostly focused on small detunings applicable to MOTs. Therefore, we attempt to verify our hypothesis by measuring the population in the ground state $F = 4$ manifold as a function of PA time. We expect excitation of the quasi-molecule to cause population of the $F = 4$ manifold as the excited potential may decay into both $F = 3$ and $F = 4$ in this case.

The results of our investigation of the $F = 4$ population are presented in the bottom panel of Fig. 7.6. The $F = 4$ number is measured by performing absorption imaging without repump present, so only atoms in the $F = 4$ manifold are detected by the resonant probe light. The number is normalised to the total Cs number which we have measured as a function of time in the upper panel Fig. 7.6. We see that although the total Cs atom number is decaying quickly, the relative number of atoms pumped into in the $F = 4$ manifold is rising quickly. This seems to verify our hypothesis, as the accumulation of population in the higher energy $F = 4$ state may only occur by the

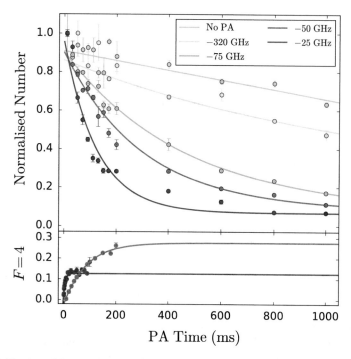

Fig. 7.6 Number of Cs atoms as a function of PA pulse time for a PA power of 125 μW. Upper: We explore a number of different detunings from the Cs D_1 line. The number is normalised to the total Cs atom number. The solid lines are exponential fits to the data. Lower: Number of Cs atoms in the $F = 4$ manifold as a function of PA pulse time. The number is normalised to the total Cs atom number measured in the experiment in the upper part of the figure

radiative decay from the quasi-molecular $6P_{1/2}$ state. However, further verification of this would require additional measurements which are tangential to our main goal of photoassociation of CsYb. Nevertheless, it is important to note that a significant fraction of Cs atoms populate the $F = 4$ state. As these atoms in the $F = 4$ manifold will contribute a background to any PA loss signals, we selectively remove these atoms by applying a pulse of probe light after the PA pulse and before absorption imaging.

In Fig. 7.7 we compare the PA spectra observed in the dipole trap to the spectra observed in a MOT. As the PA features in the MOT originate from $F = 4 + F = 4$ collisions, the internal energy of the collision pair is greater and excitation to the 0_u^+ potential requires $2 \times \hbar\Delta_{HF} \approx h \times 18.384$ GHz less energy than excitation from the $F = 3 + F = 3$ collisions present in the dipole trap. Therefore, to show both resonances in the same figure, the detuning of the MOT features are offset by twice the Cs ground state hyperfine splitting.

The most striking difference between the two techniques is the difference in widths of the features. By using a common scale for the data it is hard to see the lineshape of the dipole trap features but in the inset the Lorentzian lineshape of the $v = 134$ line is clear. The width of the feature is commensurate with twice the atomic

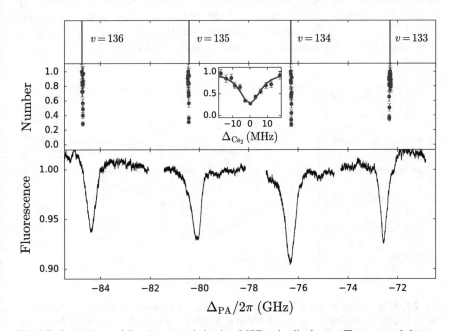

Fig. 7.7 Comparison of Cs_2 photoassociation in a MOT and a dipole trap. The top panel shows a selection of Cs_2 lines from the 0_u^+ potential observed in the dipole trap. The Cs number is plotted as a function of detuning from the Cs D_1 $F = 3 \rightarrow F' = 4$ transition. The inset shows an enhanced view of the $v = 134$ feature. The bottom panel shows the same vibrational levels observed in a MOT. The MOT fluorescence is plotted as a function of detuning from the Cs D_1 $F = 4 \rightarrow F' = 4$ transition minus twice the Cs ground state hyperfine splitting (see text)

linewidth, $2 \times \Gamma_{D_1} = 2\pi \times 9.1$ MHz. The MOT linewidths are significantly broader than the features observed in the dipole trap, this is partly due to the scan rate of the laser (2 GHz/min) and partly due to power broadening of the lines due to the large intensity of the PA light $I_{PA} = 250$ W/cm². The depth of Cs dipole trap features is 10× greater than the depth of the MOT features, even for a moderate PA intensity of $I_{PA} = 0.4$ W/cm². The larger depth of the dipole trap features is due to the increased density and the much longer interrogation times available.

7.3.3 Feshbach-Enhanced Photoassociation

Cs atoms in the $|F = 3, m_F = +3\rangle$ state possess a rich Feshbach structure [5, 16], which at low magnetic fields is dominated by the strong resonance at -12 G. The rich Feshbach structure provides a broad tunability of the scattering length that we exploit in Chap. 5 to efficiently evaporate Cs. Here, we exploit the broad tunability of the Cs scattering length to modify the Cs₂ photoassociation rate. The results of the investigation are presented in Fig. 7.8. We sample the Cs₂ photoassociation rate at a variety of magnetic fields corresponding to a range of positive and negative scattering lengths. We take care to avoid magnetic fields near narrow Feshbach resonances at 14.4, 15.1 and 19.9 G. We see that the photoassociation rate is strongly suppressed as a_{Cs} approaches zero, with the behaviour symmetric about this point.

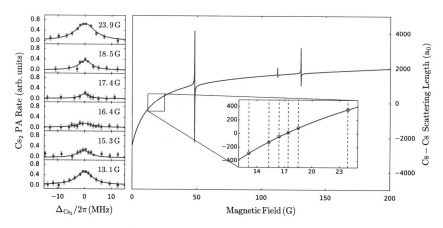

Fig. 7.8 Modification of Cs₂ photoassociation rate using a Feshbach resonance. The left panel shows Cs₂ photoassociation rates as a function of detuning from the 0_u^+ $v = 136$ line for varying magnetic field strengths. The right panel shows the Cs scattering length as a function of magnetic field. The inset shows the range of scattering lengths covered in the experiment, with the vertical dotted lines indicating the exact magnetic fields used. The scattering lengths are from Ref. [5]. Feshbach resonances at 14.4, 15.1 and 19.9 G do not appear due to the resolution of the magnetic field steps

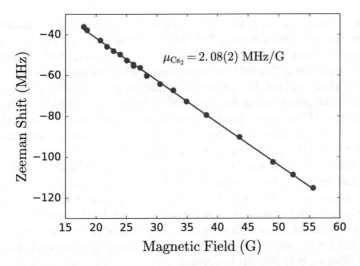

Fig. 7.9 Zeeman shift of the the 0_u^+ $v = 136$ Cs$_2$ photoassociation line. Shift of the photoassociation line centre from the zero field value versus magnetic field strength

This effect is well understood in the context of Feshbach-Optimized Photoassociation (FOPA) [41, 63, 82] and arises due to a phase shift of the scattering wavefunction. The phase shift of the scattering wavefunction is controlled by tuning the magnetic field around a Feshbach resonance and leads to a change in the FCF of the photoassociation transition as the overlap of the ground and excited state wavefunctions is modified. This effect is typically used to enhance the PA rate of a transition for deeply bound levels [47]. However, in our experiment we utilise this effect to suppress the Cs$_2$ PA rate when searching for Cs*Yb PA lines in the next section. This is not expected to modify the CsYb PA rate as the predicted Feshbach resonances in this system are very sparse and narrow [13] such that we do not expect the phase of the scattering wavefunction to vary with magnetic field strength.

The Zeeman shift of the 0_u^+ $v = 136$ photoassociation line centre as a function of magnetic field strength is shown in Fig. 7.9. The $\Omega = 0$ states do not possess hyperfine structure to first order [38], so the Zeeman shift of the transition under consideration is dominated by the energy shift of the atomic collisional state. The magnetic moment extracted from a linear fit to the data is $\mu_{Cs_2} = 2.08(2)$ MHz/G. This is consistent with the predicted value of $\mu = 2.1$ MHz/G for two colliding Cs atoms with $M_F = +6$.

7.4   Photoassociation of CsYb

In this section we begin our search for the first CsYb photoassociation lines. Observation of heteronuclear photoassociation is more challenging than photoassociation in the homonuclear case due to the difference in the asymptotic forms of the electronically excited potentials. In heteronuclear systems the electronically excited potentials no longer have the dipole-dipole R^{-3} asymptotic interaction of homonuclear systems. They instead posses van der Waals R^{-6} asymptotic behaviour, similar to the ground state potential. The R^{-6} interaction is confined to a shorter range than the R^{-3} case which drastically reduces the strength of transitions to electronically excited states from the initial atomic scattering state. This is because the Condon points for the R^{-6} vibrational levels are at shorter interatomic distances which reduces the Franck-Condon overlap. In addition, the density of states in a heteronuclear system is much lower due to the R^{-6} form. The sparsity of spectral features is further compounded in the alkali-Yb case as there is only one molecular potential that asymptotically approaches the Cs D_1 line, compared with the 3 attractive potentials that approach the Cs D_1 line in the Cs_2 case. Fortunately, CsYb possess a large reduced mass, facilitating a higher density of bound states near threshold. Nevertheless, for the maximum detuning explored in the next section, 500 GHz, there are only 20 CsYb vibrational levels predicted to exist. In this same range of detunings below the Cs D_1 line there are over 300 Cs_2 levels [64].

The photoassociation process in the CsYb case considered here is illustrated in Fig. 7.10. The energy difference between the atomic transition $\hbar\omega_0$ and the photoassociation light $\hbar\omega_1$ gives the binding energy of the vibrational level E_b in the electronically excited molecular state

$$E_b = \hbar\omega_0 - \hbar\omega_1. \tag{7.8}$$

Here, we investigate the binding energies below the $Cs(^2P_{1/2}) + Yb(^1S_0)$ asymptote. The electronic state at this threshold is designated 2(1/2) to indicate that it is the second (first excited) state with total electronic angular momentum $\Omega = 1/2$ about the internuclear axis. It correlates at short range with the $1\,^2\Pi_{1/2}$ electronic state in Hund's case (a) notation [58], but at long range the $^2\Pi_{1/2}$ and $^2\Sigma_{1/2}$ states are strongly mixed by spin-orbit coupling (see Fig. 7.1).

7.4.1 CsYb Photoassociation Routine

The preparation of the CsYb mixture in the optical dipole trap is near-identical to the routine described in Sect. 6.1 in the thermalisation measurements. Yb atoms are prepared in the trap first by loading the dipole trap from the Yb MOT using the same experimental conditions described in detail previously. The Yb atoms are then evaporatively cooled by reducing the power in the dimple beams until the Yb atoms reach a temperature of $T_{Yb} = 800$ nK in a trap with a depth of $U_{Yb} = 5\,\mu K$. There are typically $N_{Yb} = 1 \times 10^6$ ^{174}Yb atoms at this stage. The Yb atoms are stored in the dimple while the Cs MOT is loaded, compressed and cooled using optical molasses. Then, DRSC is applied to the Cs atoms for 8 ms, cooling them down to $T_{Cs} = 2\,\mu K$. During the DRSC stage some of the Cs atoms are loaded into the dimple trap as in the thermalisation measurements. We can increase the number of

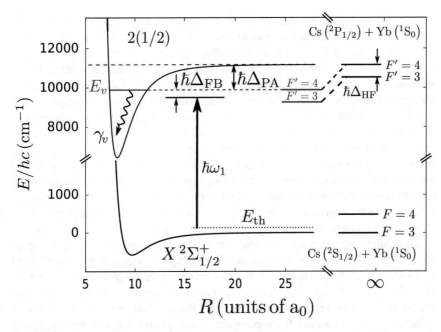

Fig. 7.10 One-photon photoassociation. When $\hbar\omega_1 = E_v$ ($\hbar\Delta_{FB} = 0$) a pair of colliding ground-state Cs and Yb atoms are associated to form a CsYb molecule in a rovibrational level of the electronically excited 2(1/2) molecular potential. The molecular curves plotted here are adapted from Ref. [58]. The hyperfine splitting shown on the right is not to scale

Cs atoms in the dimple trap by transferring the DRSC cooled cloud into the reservoir trap and holding the reservoir-trapped Cs atoms over the dimple trap containing the Cs+Yb mixture. However, for Yb isotopes which are harder to cool or with a large interspecies interaction with Cs, this step is skipped due to the large Yb loss from collisions with the high number of hotter Cs atoms trapped in the reservoir.

At the end of these stages the dimple trap typically contains a mixture of 8×10^5 ^{174}Yb atoms at $T_{Yb} = 1\,\mu$K in their spin-singlet ground state $^1S_0$ and 7×10^4 Cs atoms at $T_{Cs} = 6\,\mu$K in their absolute ground state $^2S_{1/2}\,|F = 3, m_F = +3\rangle$. This mixture is the starting point for the photoassociation measurements presented in the following section.

To search for photoassociation features we apply a pulse of photoassociation light for a variable time of 50–300 ms. We then apply an imaging pulse to remove any Cs atoms pumped into the $F = 4$ state before releasing the atoms by turning off the dimple trap and imaging the two atomic species using standard resonant absorption imaging.

Unfortunately, due to the large difference in polarisability of the two species we are only able to create a mixture of Cs+Yb with a large number imbalance in favour of Yb. This necessitates the detection of atomic Cs loss for the identification of a photoassociation line as detection using the Yb number would require us to detect a <5% change in number. Resonant loss of Cs atoms is also an indicator of a Cs_2 photoassociation line, which are spectrally dense below the D_1 line, as we have seen. This turns our search for a CsYb resonance from 'searching for a needle in a haystack' to 'searching for a specific needle in a haystack littered with needles'. Whereby we must check each feature against an extensive survey of vibrational levels conducted by Pichler and co-workers [64]. A further problem is that these measurements performed in a MOT are much lower resolution than our scans and there is the possibility that CsYb features occur close to a Cs_2 resonance. Alternatively, we could reduce the number of Yb atoms to create an almost even mixture of the two species, allowing unambiguous detection of the resonances in the Yb number, however, this would decrease the photoassociation rate by a factor of >100 due to the decrease in the interspecies density. Instead, we verify the observation of a CsYb resonance by repeating the measurement in the absence of Yb. If the resonance disappears in the absence of Yb atoms, we can be sure the feature resulted from CsYb photoassociation.

Fortunately, we can filter out a large number of 'False-positives' using the technique explored in the previous section. Conducting our search at a magnetic field of 16.4(2) G, we eliminate Cs_2 PA features arising from the $|F = 3, m_F = +3\rangle + |F = 3, m_F = +3\rangle$ scattering state. While this technique is successful, we note that we do still observe very weak features, most likely arising from $F = 4 + F = 3$ collisions (which occur due to the pumping of atoms into the $F = 4$ state by the PA light).

7.4.2 $Cs^{174}Yb$ Photoassociation Lines

We observe photoassociation lines with detunings from 17 GHz to 500 GHz. Lines with smaller detunings are hard to observe because of strong off-resonant scattering from the atomic transition and the large density of Cs_2 lines. A typical CsYb spectrum is shown in Fig. 7.11 as a function of the detuning Δ_{FB} from the free-bound transition. When the frequency of the PA laser is tuned into resonance with a CsYb line, we observe a loss of Cs atoms due to the formation of Cs^*Yb molecules. We verify that the detected features are CsYb resonances by repeating the scan in the absence of Yb. To keep the density and temperature of the Cs atoms comparable to the measurement taken with Yb, we simply remove Yb from the trap with a pulse of light resonant with the $^1S_0 \rightarrow ^1P_1$ transition immediately before the sample is illuminated by the PA light. The disappearance of the feature in the absence of Yb (red trace in Fig. 7.11) confirms the existence of a CsYb PA resonance.

The typical width of the observed lines varies from 5 to 10 MHz ($\Gamma_{D1}/2\pi = 4.56$ MHz). We observe larger widths for more deeply bound lines due to power broadening as the intensity of the PA pulse is increased to maintain a good SNR for vibrational levels with a lower Franck-Condon factor.

For all $^{133}Cs^{174}Yb$ vibrational levels we observe a second PA feature which is red detuned by approximately the hyperfine splitting of the Cs $6P_{1/2}$ level. For the weakly-bound vibrational states investigated here, the Cs^*Yb molecules inherit the

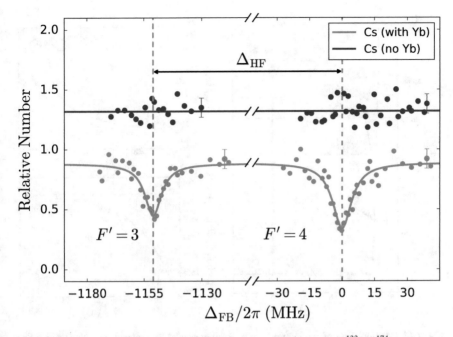

Fig. 7.11 Observation of the photoassociation resonance for $n' = -11$ of $^{133}\mathrm{Cs}^{*174}\mathrm{Yb}$. Relative number of Cs atoms remaining after a 300 ms pulse of PA light versus detuning from the $F' = 4$ line (Δ_{FB}). The green (red) trace shows the photoassociation spectra of Cs with (without) the presence of Yb in the dipole trap. The red (Cs only) trace has been offset for clarity. The statistical error in the atom number is shown by the error bars on the right hand side. The dashed green line shows the centres of the CsYb PA resonance for each hyperfine component

properties of the two free atoms; as such we identify the two lines by the quantum numbers $F' = 4$ and $F' = 3$ corresponding to the hyperfine structure in the excited state of Cs. The rovibrational levels are best described by the classic form of Hund's case (e) introduced by Mulliken [60], in which the total atomic angular momentum (F' here) couples to the rotational angular momentum ℓ to form a resultant $\mathcal{F}'$. This uncommon coupling case was first observed for HeKr$^+$ [15] and has also been found in RbYb [14, 62]. In Hund's case (e) the rotational progression is described by $E_{\mathrm{rot}} = B_{\mathrm{rot}}\ell(\ell + 1)$. In our case, the temperature of the initial mixture is well below the p-wave barrier height of 40 μK, so we do not populate higher rotational levels (in the absence of shape resonances), therefore, all observed features have $\ell = 0$.

Hyperfine Splitting of Photoassociation Lines

The CsYb spectra display hyperfine structure associated with the Cs atom, as shown in Fig. 7.11. We present the measured hyperfine splitting for all the observed $^{133}\mathrm{Cs}^{174}\mathrm{Yb}$ levels in Table 7.1 and we illustrate the dependence of the hyperfine coupling on internuclear distance in Fig. 7.12. We tune the internuclear distance between Cs and Yb by creating Cs*Yb molecules in different vibrational levels with an array of binding energies. We approximate the effective internuclear distance R_{eff} for each

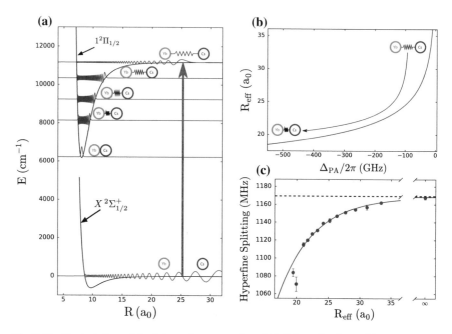

Fig. 7.12 Manipulation of the internuclear separation of Cs*Yb molecules. **a** Photoassociation of Cs*Yb molecules with variable internuclear distance. The black lines show the ground and excited state of the CsYb molecular potential in Lennard-Jones potential form. The red lines show the vibrational (collisional) wavefunctions calculated by solving the radial Schrödinger equation for the excited (ground) state potential. **b** Detuning of the photoassociation line plotted against the effective internuclear distance, R_{eff}, calculated from the molecular potentials. **c** Measured hyperfine splitting as a function of the effective internuclear distance, R_{eff}. The horizontal dashed line shows the Cs atomic hyperfine splitting of the $m_F = +3$ levels in the $6P_{1/2}$ state at a magnetic field of 2.2(2) G [72, 83]. The solid line is an exponential fit

transition as the Condon point, where the transition energy is equal to the spacing between the two curves.

The points show the measured hyperfine splitting of the $F' = 4, m_F' = 3$ and $F' = 3, m_F' = 3$ sublevels of each vibrational level and the solid red line shows an exponential fit through the points. The measured atomic value is in agreement with the literature value of 1169.272(81) MHz [83] for the hyperfine splitting of the $m_F = +3$ levels in a 2.2 G magnetic field. The strength of the Cs hyperfine coupling decreases as the binding energy increases and the internuclear separation reduces because the electronic wave function of the Cs atom is perturbed by the presence of the closed-shell Yb atom [90]. This is analogous to the effect predicted to produce Mechanism I Feshbach resonances in collisions between ground state Cs and Yb atoms.

The deepest bound level $n' = -20$ exhibits a hyperfine splitting of $\Delta_{HF} = 1084(6)$ MHz, a reduction of almost 100 MHz from the atomic value. The hyperfine splitting of $n' = -19$ is even smaller than this, but this may be due to mixing

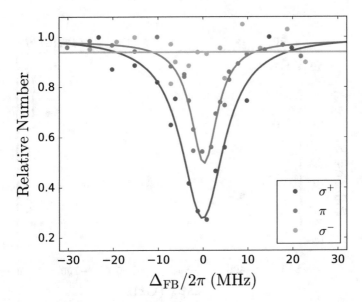

Fig. 7.13 Polarisation dependence of a ^{133}Cs^{174}Yb photoassociation resonance. Relative number of Cs atoms remaining in trap after a 300 ms pulse of photoassociation light versus detuning from the $F' = 4$ CsYb feature (Δ_{FB}) for light driving σ_+ (red), π (blue) and σ_- (green) transitions

of vibronic states in other electronic states causing a modification of the coupling, as has been observed for the case of rotational coupling in homonuclear PA [4, 66, 67]. The changing coupling strength highlights the inefficacy of labelling the angular momenta coupling using strict Hund's case definitions. It is evident that for shorter internuclear distances the Cs*Yb molecule will no longer be well described by Hund's case (e) and will transition into a more intermediate coupling regime.

Line Strengths

In the absence of rotation, the transition strengths to weakly bound vibrational levels are dictated purely by the atomic dipole matrix elements and the Franck-Condon factor.

In Fig. 7.13 we study the strength of the $F' = 4$, $n' = -11$ line for a variety of polarisations. The $\sigma^{+/-}$ transitions are driven with circularly polarised light and a 2.2 G magnetic bias field oriented along the propagation axis of the PA light. The π transition is driven with vertically polarised light and a 2.2 G bias field in the vertical direction. As expected the circularly polarised light driving σ^+ transitions produces the strongest loss feature, as the matrix element for this transition is $\sqrt{7/12}$ in comparison to $\sqrt{7/48}$ for the linearly polarised light [72]. The Cs loss due the circularly polarised light driving σ^- transitions is not observed due to the matrix element of the transition, $\sqrt{1/48}$, being $1/\sqrt{28}$ of the σ^+ matrix element. The Cs number is plotted relative to the total Cs number off resonance for each polarisation as the background Cs number is also strongly dependent on the polarisation. This is

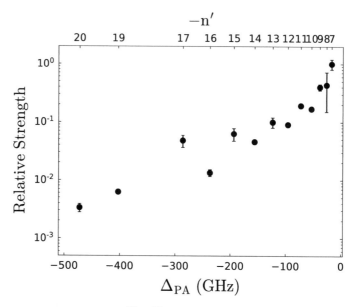

Fig. 7.14 Line strengths of observed ^{133}Cs^{174}Yb photoassociation lines. The loss rate is normalised to that of $n' = -7$ and plotted as a function of binding energy

because both the molecular photoassociation signal and the off-resonant loss of Cs atoms are dictated by the same dipole matrix elements.

With the polarisation of the PA light fixed to drive σ^+ transitions, we study the strength of transitions to different vibrational levels by observing the loss of Cs atoms versus intensity of PA light with the light on a CsYb resonance. We observe an exponential decay of the Cs atom number as a function of intensity of the PA light. This is partially due to the light-assisted collisions investigated earlier but also due to the photoassociation of CsYb. We distinguish between these loss rates by measuring the decay in the presence of Yb (both loss processes present) and without Yb present (only light-assisted losses). The difference between these loss rates is the PA loss rate.

The photoassociation decay constant extracted from the exponential fit for $F' = 4$ lines is normalised to that of the $n' = -7$ transition and given in Table 7.1 alongside the binding energy measurements. The relative strengths of the transitions are also displayed in Fig. 7.14. As the same polarisation is used for the observation of all the vibrational levels, the variations in the strength of the transitions are due to variations in the FCF. The oscillations in Fig. 7.14 are due to the varying overlap of the oscillating vibrational and scattering wavefunctions.

7.4.3 CsYb Binding Energies

Table 7.1 lists the detunings Δ_{PA} of all observed photoassociation lines for $^{133}Cs^{174}Yb$. The line frequencies are measured relative to the Cs $6S_{1/2}, F = 3 \rightarrow 6P_{1/2}, F' = 4$ atomic transition, using the difference in EOM modulation frequencies and number of cavity FSRs between the PA transition and the atomic transition, as outlined earlier. The atomic and molecular transitions are measured under the same trapping conditions; for the near-threshold states considered here, the AC Stark shifts due to the trapping light are essentially identical to those for the atoms. These measurements were performed at a magnetic field of 2.2(2) G to reduce uncertainty caused by the Zeeman shift of the molecular state when measuring the hyperfine splitting. The uncertainty due to the stabilisation of the cavity length is the dominant source of uncertainty for the majority of the measured binding energies. The exception is the $n' = -19$ line, where the observed FWHM linewidth of 130(10) MHz leads to a larger uncertainty in determining the line centre.

LeRoy-Bernstein Analysis

We obtain the binding energies E_b as the absolute value of the weighted means of the detunings for $F' = 3$ and 4, corrected for magnetic field. These are included in Table 7.1 and shown in Fig. 7.15. These measured binding energies are very small compared to the depth of the potential (≈ 200 THz), so the positions of the vibra-

Table 7.1 Measured detunings of photoassociation lines to vibrational levels of $^{133}Cs^{174}Yb$ in the 2(1/2) excited state, together with the corresponding hyperfine splittings, binding energies and line strengths. The uncertainties quoted are 1σ uncertainties [37]. The observed strengths are for the $F' = 4$ lines and are normalized to that of the strongest PA line, $n' = -7$

n'	$\Delta_{PA}/2\pi$ (GHz)	$\Delta_{HF}/2\pi$ (GHz)	E_b/h (GHz)	Normalized strength
Cs	0	1.168(2)	0	N/A
−7	−17.244(3)	1.162(1)	17.241(3)	1.0(2)
−8	−26.473(3)	1.157(3)	26.468(4)	0.4(3)
−9	−38.567(3)	1.154(1)	38.560(3)	0.40(5)
−10	−53.932(3)	1.151(1)	53.924(3)	0.17(1)
−11	−72.973(3)	1.147(1)	72.963(3)	0.19(1)
−12	−96.091(3)	1.142(2)	96.079(3)	0.091(8)
−13	−123.678(3)	1.139(1)	123.665(3)	0.10(2)
−14	−156.117(3)	1.131(1)	156.100(3)	0.045(4)
−15	−193.772(3)	1.127(1)	193.753(3)	0.06(2)
−16	−236.991(3)	1.120(1)	236.969(3)	0.013(2)
−17	−286.098(4)	1.115(2)	286.074(4)	0.05(1)
−18		————not observed————		
−19	−402.867(8)	1.071(8)	402.824(9)	0.0063(4)
−20	−472.384(6)	1.084(6)	472.347(7)	0.0033(6)

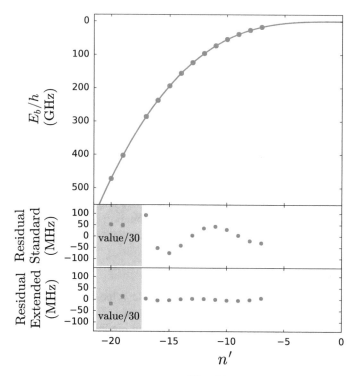

Fig. 7.15 Binding energies of vibrational levels of ^{133}Cs^{174}Yb in the 2(1/2) excited state. Top: The binding energies as a function of the vibrational quantum number counted from dissociation, n'. The solid green line shows a fit to the data using the extended Le Roy-Bernstein equation. The lower two panels compare the residuals for the fits using the standard and extended Le Roy-Bernstein equations. The error bars are much smaller than the data points

tional levels are determined principally by the long-range potential. At long range, the CsYb 2(1/2) potential is dominated by the van der Waals $n = 6$ term. These asymptotic coefficients may be extracted from PA spectra using near-dissociation expansion formulae, such as the seminal Le Roy-Bernstein (LRB) formula described in Sect. 7.1.3.

In searching for PA lines, we modelled our data using Eq. 7.3 (for $n = 6$) and used the fitted parameters to predict more deeply bound levels. This technique yielded accurate predictions for levels up to $n' = -17$, with the measured binding energies typically lying within a few hundred MHz of the predicted values. For the more deeply bound levels $n' = -19$ and $n' = -20$, the measured line frequencies were far from the extrapolated values and the $n' = -18$ level was not observed at all. The non-observation of the $n' = -18$ level may be due to a small Franck-Condon factor or that the level is located outside the range searched (± 2 GHz from the prediction) or coincided with a Cs$_2$ transition. We did not search further due to the ~ 30 s load-detection cycle associated with conducting the measurements.

The middle panel of Fig. 7.15 shows the residuals from the fit of our PA measurements to the LRB equation. The $n' = -19$ and $n' = -20$ levels are outliers and so not included in any of our fits. It is clear from the residuals that the standard LRB equation does not fully describe our measured PA spectra. The structure of the residuals suggests that a model including higher-order terms would give a better fit to the results. Indeed, the more strongly bound levels with binding energies around 300 GHz are deep enough to be sensitive to the non-asymptotic, short-range character of the potential for our measurement precision.

To model the PA spectra better, we also fit them using an extended version of the LRB equation [20]

$$
E_b \approx \left(\frac{v_D - v}{B_n} \right)^{\frac{1}{1-\beta}}
$$

$$
\times \left[1 - \frac{1}{1-\beta} \frac{1}{v_D - v} \left(\gamma \left(\frac{v_D - v}{B_n} \right)^{\frac{1}{1-\beta}} \right. \right.
$$

$$
+ \frac{\sqrt{2\mu}}{2\pi\hbar} \frac{C_n^{\delta-1/2}}{n} \frac{C_m}{C_n} \frac{1}{1-\delta} \left(\frac{v_D - v}{B_n} \right)^{\frac{1-\delta}{1-\beta}}
$$

$$
\times \left. \left. \begin{cases} \beta B(\beta, \frac{1}{2}) & \text{if } \delta < 0 \\ \frac{1}{2} \ln \left(\frac{v_D - v}{B_n} \right)^{\frac{1}{1-\beta}} & \text{if } \delta = 0 \\ (\delta - \frac{1}{2}) B(\delta, \frac{1}{2}) & \text{if } \delta > 0 \end{cases} \right) \right], \tag{7.9}
$$

where $\beta = (n + 2 - 2l)/2n$ and $\delta = \beta + (n - m)/n$. The extended version allows the inclusion of higher order dispersion coefficients (we use $m = 8$) in addition to a parameter γ which accounts for the non-asymptotic, short-range character of the potential. The bottom panel of Fig. 7.15 shows the residuals of the fit to the extended LRB equation. The inclusion of the extra terms significantly improves the fit to the data. The reduced chi-squared of the extended fit is $\chi_\nu^2 = 1.8$, a much better fit in comparison to the standard LRB which gives a reduced chi-squared of $\chi_\nu^2 = 260$. The best-fit parameters for the extended fit are $C_6 = 10.1(1) \times 10^3 \, E_h a_0^6$, $C_8 = 4.9(2) \times 10^6 \, E_h a_0^8$, $v_{\text{frac}} = 0.695(6)$ and $\gamma^{-1} = h \times 3.4(1) \times 10^2$ GHz. Despite the improvement in the fit due to the inclusion of the higher order terms in the improved LRB, a more complete description of the experimental data requires further higher-order terms (C_{10}, C_{12}, etc.) and the inclusion of coupling between potentials. However, no NDEs exist for the inclusion of further dispersion coefficients, so further analysis would require numerical solution of a coupled channel model of the potential which is beyond the scope of this investigation.

When fitting to either model, the residuals for $n' = -19$ and -20 are over 30 times larger than that of the other levels. These levels may be perturbed by mixing with

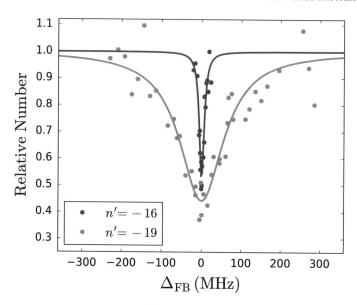

Fig. 7.16 Anomalous broadening of the ^{133}Cs^{174}Yb $n' = -19$ line. The plot shows a PA spectrum of the $n' = -16$ (red) and $n' = -19$ (green) ^{133}Cs^{174}Yb lines recorded with the same PA intensity $I_{PA} = 34\,\text{W/cm}^2$

vibrational levels in a different electronic state [11]. The shift could also be caused by the broadband dipole trapping light coupling to a higher electronic state. The $n' = -19$ line is extremely broad in comparison to other observed lines; it has a FWHM of 130(10) MHz, over eight times the linewidth of $n' = -16$ (FWHM = 15(2) MHz) at the same light intensity. The disparity in the spectra of these two lines is illustrated in Fig. 7.16. We have not been able to observe any levels beyond $n' = -20$, although we have searched a moderate ± 1 GHz range around the predicted positions. As can be seen from the residuals for the deepest observed states in Fig. 7.15, the disagreement with the LRB fit results in an increasingly large search space, which is very time consuming to explore.

7.4.4 Photoassociation of Other CsYb Isotopologs

One of the distinct advantages of using Yb in an atomic mixture is the number of abundant stable isotopes, both bosonic and fermionic, that can be trapped and cooled to ultracold temperatures [30, 31, 74–77]. In the context of this work, measurements using a range of CsYb isotopologs can be combined to put tighter constraints on the 2(1/2) electronic state. Here we apply the method of mass scaling to determine the maximum number of bound states supported by the 2(1/2) state. Within the Born-Oppenheimer approximation, the interaction potential is mass-independent but the positions of vibrational levels depend on the reduced mass.

Table 7.2 Measured detunings of photoassociation lines to vibrational levels of different iso-topologs of ^{133}CsYb in the 2(1/2) excited state. The detunings are for the $F' = 4$ sublevel. The residuals given are from the extended LRB model with $N_{\text{vib}} = 154$ or $N_{\text{vib}} = 155$

Yb isotope	n'	$\Delta_{\text{PA}}/2\pi$ (GHz)	Residual (MHz)	
			$N_{\text{vib}} = 154$	$N_{\text{vib}} = 155$
173	−9	−36.117(3)	10	−6
173	−10	−50.877(3)	10	−11
173	−11	−69.246(3)	15	−10
173	−12	−91.633(3)	19	−12
173	−13	−118.427(3)	23	−13
173	−14	−150.014(3)	29	−14
173	−15	−186.762(3)	38	−11
172	−8	−22.740(5)	14	−10
172	−11	−65.614(4)	21	−28
172	−13	−113.258(4)	45	−26
170	−12	−103.338(3)	−15	−150
170	−14	−166.489(3)	−55	−241

In WKB quantization, the non-integer quantum number at dissociation, $v_D = v_{\text{max}} + v_{\text{frac}}$, is given by $v_D = \Phi/\pi - 1/2$, where Φ is the phase integral

$$\Phi = \int_{R_{\text{in}}}^{\infty} \sqrt{\frac{2\mu_r}{\hbar^2} V(R)} \, dR. \tag{7.10}$$

Here R_{in} is the location of the inner classical turning point, μ_r is the reduced mass and $V(R)$ is the interaction potential. The dependence on μ_r allows us to determine the number of bound states $N_{\text{vib}} = v_{\text{max}} + 1$ by comparing binding energies for different isotopologs.

The measured binding energies of ^{133}Cs^{173}Yb, ^{133}Cs^{172}Yb and ^{133}Cs^{170}Yb are tabulated in Table 7.2. The routines used to obtain PA spectra for these isotopologs are similar to that presented for ^{133}Cs^{174}Yb, with the only significant difference in the preparation of the ultracold Yb sample. Slight changes are required to the MOT, dimple trap loading and evaporative cooling routines to address the differ-ent requirements of each Yb isotope due to variations in abundance, intraspecies scattering length and hyperfine structure (for fermionic ^{173}Yb). The ^{133}Cs^{173}Yb and ^{133}Cs^{170}Yb measurements take place in identical trapping conditions to ^{133}Cs^{174}Yb. The initial mixture contains 3×10^5 ^{173}Yb or 4×10^5 ^{170}Yb atoms at $T_{\text{Yb}} = 1\,\mu\text{K}$ and 5×10^4 Cs atoms at $T_{\text{Cs}} = 6\,\mu\text{K}$. The large negative scattering length of ^{172}Yb ($a_{172-172} = -600\,a_0$) [43] complicates the evaporative cooling of Yb; we therefore halt the evaporation around $T_{\text{Yb}} = 4\,\mu K$ to prevent a substantial loss of Yb atoms due to 3-body inelastic collisions. PA for ^{133}Cs^{172}Yb is performed on a mixture of 5×10^5 ^{172}Yb atoms at $T_{\text{Yb}} = 4\,\mu\text{K}$ and 7×10^4 Cs atoms at $T_{\text{Cs}} = 12\,\mu\text{K}$. In this new trapping arrangement the Yb (Cs) trap frequencies are 380 (1100) Hz radially and 80 (240) Hz axially.

To determine N_{vib} from the measured binding energies of the four isotopologs we use a mass-scaled version of the extended LRB model. The values of C_6 and C_8 are the same for all isotopologs. However, v_D is proportional to $\sqrt{\mu_r}$, and so v_{frac} varies between isotopologs. γ is also proportional to $\sqrt{\mu_r}$ [20], but this variation is much less important than that for v_{frac}. For a chosen value of N_{vib}, we can use the parameters fitted to the $^{133}\text{Cs}^{174}\text{Yb}$ binding energies to predict binding energies for the other isotopologs and calculate χ_ν^2. It is possible to refit the parameters with multiple isotopologs, but this makes little quantitative difference and produces the same qualitative conclusions.

The binding energies for $^{133}\text{Cs}^{172}\text{Yb}$ and $^{133}\text{Cs}^{173}\text{Yb}$ are well predicted by the parameters obtained for $^{133}\text{Cs}^{174}\text{Yb}$ with $N_{\text{vib}} = 155$, giving $\chi_\nu^2 = 12$. This compares with $\chi_\nu^2 = 40$ and 158 for $N_{\text{vib}} = 154$ and 156 respectively. However, including $^{133}\text{Cs}^{170}\text{Yb}$ gives $\chi_\nu^2 = 36$ for $N_{\text{vib}} = 154$ and 322 for $N_{\text{vib}} = 155$.

We attempt to quantitatively evaluate the statistical uncertainty in N_{vib} using a Bayesian approach, as the large reduced chi-squared values mean that the covariance matrix does not yield a good approximation of the uncertainty in N_{vib}. The use of Bayesian inference is more applicable in our case as it is not extremely sensitive to the assumption that the model is a good fit to the data [71]. We sample the posterior probability density function using Goodman & Weare's affine invariant Markov chain Monte Carlo ensemble sampler [33] implemented using emcee [29]. We calculate the 3σ confidence (credible) interval from the quantiles of the marginalised distribution. From this approach we find $N_{\text{vib}} = 154$ with 3σ confidence for the inclusion of all the data but $N_{\text{vib}} = 155$ with 3σ confidence with the exclusion of $^{133}\text{Cs}^{170}\text{Yb}$ data. It thus appears that the results for the different isotopologs are inconsistent with a single-potential model; the deviations are outside the experimental errors and appear non-statistical.

It is possible that the lines for one or more isotopes are affected by an isotope-dependent perturbation, most likely due to a level of the 3(1/2) electronic state that dissociates to the $6\,^2P_{3/2}$ state of Cs. Such a perturbation is not encapsulated in our model and characterizing it would require extensive further work. Nevertheless, we can conclude that the number of bound states supported by the 2(1/2) potential is either 154 or 155. This is within 10% of the 145 bound states predicted for this potential by Meniailava and Shundalau [58].

References

1. Abraham ERI, McAlexander WI, Gerton JM, Hulet RG, Côté R, Dalgarno A (1996) Singlet *s*-wave scattering lengths of ^{6}Li and ^{7}Li. Phys Rev A 53:R3713–R3715. https://doi.org/10.1103/PhysRevA.53.R3713
2. Aikawa K, Akamatsu D, Hayashi M, Oasa K, Kobayashi J, Naidon P, Kishimoto T, Ueda M, Inouye S (2010) Coherent transfer of photoassociated molecules into the rovibrational ground state. Phys Rev Lett 105(20):203,001. https://doi.org/10.1103/PhysRevLett.105.203001

3. Altaf A, Dutta S, Lorenz J, Pérez-Ríos J, Chen YP, Elliott DS (2015) Formation of ultracold ^{7}Li ^{85}Rb molecules in the lowest triplet electronic state by photoassociation and their detection by ionization spectroscopy. J Chem Phys 142(11):114,310. https://doi.org/10.1063/1.4914917

4. Bergeman T, Qi J, Wang D, Huang Y, Pechkis HK, Eyler EE, Gould PL, Stwalley WC, Cline RA, Miller JD, Heinzen DJ (2006) Photoassociation of ^{85}Rb atoms into 0_u^+ states near the 5S+5P atomic limits. J Phys B At Mol Opt Phys 39(19):S813. https://doi.org/10.1088/0953-4075/39/19/s01

5. Berninger M, Zenesini A, Huang B, Harm W, Nägerl HC, Ferlaino F, Grimm R, Julienne PS, Hutson JM (2013) Feshbach resonances, weakly bound molecular states, and coupled-channel potentials for cesium at high magnetic fields. Phys Rev A 87(3):032,517. https://doi.org/10.1103/physreva.87.032517

6. Black ED (2001) An introduction to Pound-Drever-Hall laser frequency stabilization. Am J Phys 69(1):79–87. https://doi.org/10.1119/1.1286663

7. Boesten HMJM, Tsai CC, Verhaar BJ, Heinzen DJ (1996) Observation of a shape resonance in cold-atom scattering by pulsed photoassociation. Phys Rev Lett 77(26):5194–5197. https://doi.org/10.1103/physrevlett.77.5194

8. Bohn JL, Julienne PS (1999) Semianalytic theory of laser-assisted resonant cold collisions. Phys Rev A 60:414–425. https://doi.org/10.1103/PhysRevA.60.414

9. Boisseau C, Audouard E, Vigué J, Julienne PS (2000) Reflection approximation in photoassociation spectroscopy. Phys Rev A 62(052):705. https://doi.org/10.1103/PhysRevA.62.052705

10. Borkowski M (2018) Optical lattice clocks with weakly bound molecules. Phys Rev Lett 120(8):083,202. https://doi.org/10.1103/physrevlett.120.083202

11. Borkowski M, Morzyński P, Ciuryło R, Julienne PS, Yan M, DeSalvo BJ, Killian TC (2014) Mass scaling and nonadiabatic effects in photoassociation spectroscopy of ultracold strontium atoms. Phys Rev A 90(032):713. https://doi.org/10.1103/PhysRevA.90.032713

12. Brown JM, Carrington A (2003) Rotational spectroscopy of diatomic molecules. Cambridge University Press. https://doi.org/10.1017/cbo9780511814808

13. Brue DA, Hutson JM (2013) Prospects of forming ultracold molecules in $^2\Sigma$ states by magnetoassociation of alkali-metal atoms with Yb. Phys Rev A 87(5):052,709. https://doi.org/10.1103/physreva.87.052709

14. Bruni C, Görlitz A (2016) Observation of hyperfine interaction in photoassociation spectra of ultracold RbYb. Phys Rev A 94(2):022,503. https://doi.org/10.1103/physreva.94.022503

15. Carrington A, Pyne CH, Shaw AM, Taylor SM, Hutson JM, Law MM (1996) Microwave spectroscopy and interaction potential of the long-range He$\cdots$Kr$^+$ ion: an example of Hund's case (e). J Chem Phys 105(19):8602–8614. https://doi.org/10.1063/1.472999

16. Chin C, Vuletić V, Kerman AJ, Chu S, Tiesinga E, Leo PJ, Williams CJ (2004) Precision Feshbach spectroscopy of ultracold Cs$_2$. Phys Rev A 70(3):032,701. https://doi.org/10.1103/physreva.70.032701

17. Chin C, Flambaum VV, Kozlov MG (2009) Ultracold molecules: new probes on the variation of fundamental constants. New J Phys 11(5):055,048. https://doi.org/10.1088/1367-2630/11/5/055048

18. Chin C, Grimm R, Julienne P, Tiesinga E (2010) Feshbach resonances in ultracold gases. Rev Mod Phys 82(2):1225. https://doi.org/10.1103/revmodphys.82.1225

19. Ciamei A, Bayerle A, Chen CC, Pasquiou B, Schreck F (2017) Efficient production of long-lived ultracold sr$_2$ molecules. Phys Rev A 96(013):406. https://doi.org/10.1103/PhysRevA.96.013406

20. Comparat D (2004) Improved LeRoy-Bernstein near-dissociation expansion formula, and prospect for photoassociation spectroscopy. J Chem Phys 120(3):1318–1329. https://doi.org/10.1063/1.1626539

21. Comparat D, Drag C, Tolra BL, Fioretti A, Pillet P, Crubellier A, Dulieu O, Masnou-Seeuws F (2000) Formation of cold Cs ground state molecules through photoassociation in the pure long-range state. Eur Phys J D 11(1):59–71. https://doi.org/10.1007/s100530070105

22. Danzl JG, Haller E, Gustavsson M, Mark MJ, Hart R, Bouloufa N, Dulieu O, Ritsch H, Nägerl HC (2008) Quantum gas of deeply bound ground state molecules. Science 321(5892):1062–1066. https://doi.org/10.1126/science.1159909

23. Danzl JG, Mark MJ, Haller E, Gustavsson M, Hart R, Aldegunde J, Hutson JM, Nägerl HC (2010) An ultracold high-density sample of rovibronic ground-state molecules in an optical lattice. Nat Phys 6(4):265–270. https://doi.org/10.1038/nphys153
24. Deiglmayr J, Grochola A, Repp M, Mörtlbauer K, Glück C, Lange J, Dulieu O, Wester R, Weidemüller M (2008) Formation of ultracold polar molecules in the rovibrational ground state. Phys Rev Lett 101(13):133,004. https://doi.org/10.1103/PhysRevLett.101.133004
25. Dion CM, Drag C, Dulieu O, Laburthe Tolra B, Masnou-Seeuws F, Pillet P (2001) Resonant coupling in the formation of ultracold ground state molecules via photoassociation. Phys Rev Lett 86:2253–2256. https://doi.org/10.1103/PhysRevLett.86.2253
26. Drever RWP, Hall JL, Kowalski FV, Hough J, Ford GM, Munley AJ, Ward H (1983) Laser phase and frequency stabilization using an optical resonator. Appl Phys B 31(2):97–105. https://doi.org/10.1007/BF00702605
27. Fioretti A, Comparat D, Crubellier A, Dulieu O, Masnou-Seeuws F, Pillet P (1998) Formation of cold Cs_2 molecules through photoassociation. Phys Rev Lett 80(20):4402. https://doi.org/10.1063/1.1302653
28. Fioretti A, Comparat D, Drag C, Amiot C, Dulieu O, Masnou-Seeuws F, Pillet P (1999) Photoassociative spectroscopy of the Cs_2 0_g^- long-range state. Eur Phys J D 5(3):389–403. https://doi.org/10.1007/s100530050271
29. Foreman-Mackey D, Hogg DW, Lang D, Goodman J (2013) Emcee: the MCMC hammer. Publ Astron Soc Pac 125(925):306–312. https://doi.org/10.1086/670067
30. Fukuhara T, Sugawa S, Takahashi Y (2007) Bose-Einstein condensation of an ytterbium isotope. Phys Rev A 76(5):051,604. https://doi.org/10.1103/PhysRevA.76.051604
31. Fukuhara T, Takasu Y, Kumakura M, Takahashi Y (2007) Degenerate Fermi gases of ytterbium. Phys Rev Lett 98(3):030,401. https://doi.org/10.1103/PhysRevLett.98.030401
32. Gardner JR, Cline RA, Miller JD, Heinzen DJ, Boesten HMJM, Verhaar BJ (1995) Collisions of doubly spin-polarized, ultracold ^{85}Rb atoms. Phys Rev Lett 74(19):3764–3767. https://doi.org/10.1103/physrevlett.74.3764
33. Goodman J, Weare J (2010) Ensemble samplers with affine invariance. Commun Appl Math Comput Sci 5(1):65–80. https://doi.org/10.2140/camcos.2010.5.65
34. Gregory PD, Molony PK, Köppinger MP, Kumar A, Ji Z, Lu B, Marchant AL, Cornish SL (2015) A simple, versatile laser system for the creation of ultracold ground state molecules. New J Phys 17(5):055,006. https://doi.org/10.1088/1367-2630/17/5/055006
35. Guttridge A, Hopkins SA, Frye MD, McFerran JJ, Hutson JM, Cornish SL (2018) Production of ultracold cs*Yb molecules by photoassociation. Phys Rev A 97(063):414. https://doi.org/10.1103/PhysRevA.97.063414
36. Herzberg G (1989) Molecular spectra and molecular structure: spectra of diatomic molecules. Van Nostrand, New York
37. Hughes IG, Hase TPA (2010) Measurements and their uncertainties. Oxford University Press
38. Jones KM, Tiesinga E, Lett PD, Julienne PS (2006) Ultracold photoassociation spectroscopy: long-range molecules and atomic scattering. Rev Mod Phys 78(2):483–535. https://doi.org/10.1103/revmodphys.78.483
39. Julienne P (1996) Cold binary atomic collisions in a light field. J Res Nat Inst Stand Technol 101(4):487. https://doi.org/10.6028/jres.101.050
40. Julienne PS, Suominen KA, Band Y (1994) Complex-potential model of collisions of laser-cooled atoms. Phys Rev A 49(5):3890–3896. https://doi.org/10.1103/physreva.49.3890
41. Junker M, Dries D, Welford C, Hitchcock J, Chen YP, Hulet RG (2008) Photoassociation of a Bose-Einstein condensate near a Feshbach resonance. Phys Rev Lett 101(6):060,406. https://doi.org/10.1103/physrevlett.101.060406
42. Kerman AJ, Sage JM, Sainis S, Bergeman T, DeMille D (2004) Production of ultracold polar RbCs* molecules via photoassociation. Phys Rev Lett 92(3):033,004. https://doi.org/10.1103/physrevlett.92.033004
43. Kitagawa M, Enomoto K, Kasa K, Takahashi Y, Ciuryło R, Naidon P, Julienne PS (2008) Two-color photoassociation spectroscopy of ytterbium atoms and the precise determinations of s-wave scattering lengths. Phys Rev A 77(1):012,719. https://doi.org/10.1103/physreva.77.012719

44. Köppinger M (2014) Creation of ultracold RbCs molecules. PhD thesis, Durham University
45. Kraft SD, Mudrich M, Staudt MU, Lange J, Dulieu O, Wester R, Weidemüller M (2005) Saturation of Cs_2 photoassociation in an optical dipole trap. Phys Rev A 71(1). https://doi.org/10.1103/physreva.71.013417
46. Krems RV (2008) Cold controlled chemistry. Phys Chem Chem Phys 10(28):4079–4092. https://doi.org/10.1039/B802322K
47. Krzyzewski SP, Akin TG, Dizikes J, Morrison MA, Abraham ERI (2015) Observation of deeply bound $^{85}Rb_2$ vibrational levels using Feshbach optimized photoassociation. Phys Rev A 92(6):062,714. https://doi.org/10.1103/physreva.92.062714
48. Le Roy RJ, Bernstein RB (1970) Dissociation energy and long-range potential of diatomic molecules from vibrational spacings of higher levels. J Chem Phys 52(8):3869–3879. https://doi.org/10.1063/1.1673585
49. Legaie R, Picken CJ, Pritchard JD (2018) Sub-kilohertz excitation lasers for quantum information processing with rydberg atoms. J Opt Soc Am B 35(4):892–898. https://doi.org/10.1364/JOSAB.35.000892
50. Lett PD, Julienne PS, Phillips WD (1995) Photoassociative spectroscopy of laser-cooled atoms. Annu Rev Phys Chem 46(1):423–452. https://doi.org/10.1146/annurev.pc.46.100195.002231
51. Li P, Liu W, Wu J, Ma J, Fan Q, Xiao L, Sun W, Jia S (2017) New observation and analysis of the ultracold Cs_2 0_u^+ and 1_g long-range states at the asymptote $6S_{1/2}+6P_{1/2}$. J Quant Spectrosc Radiat Transfer 196:176–181. https://doi.org/10.1016/j.jqsrt.2017.04.014
52. Lignier H, Fioretti A, Horchani R, Drag C, Bouloufa N, Allegrini M, Dulieu O, Pruvost L, Pillet P, Comparat D (2011) Deeply bound cold caesium molecules formed after 0_g^- resonant coupling. Phys Chem Chem Phys 13(42):18,910. https://doi.org/10.1039/c1cp21488h
53. Lisdat C, Vanhaecke N, Comparat D, Pillet P (2002) Line shape analysis of two-colour photoassociation spectra on the example of the Cs ground state. Eur Phys J D 21(3):299–309. https://doi.org/10.1140/epjd/e2002-00209-9
54. Liu W, Xu R, Wu J, Yang J, Lukashov SS, Sovkov VB, Dai X, Ma J, Xiao L, Jia S (2015) Observation and deperturbation of near-dissociation ro-vibrational structure of the Cs_2 state $0_u^+ \left(a^1 \Sigma_u^+ \sim b^3 \Pi_u \right)$ at the asymptote $6S_{1/2}+6P_{1/2}$. J Chem Phys 143(12):124,307. https://doi.org/10.1063/1.4931646
55. Ma J, Liu W, Yang J, Wu J, Sun W, Ivanov VS, Skublov AS, Sovkov VB, Dai X, Jia S (2014) New observation and combined analysis of the Cs_2 0_g^-, 0_u^+, and 1_g states at the asymptotes $6S_{1/2}+6P_{1/2}$ and $6S_{1/2}+6P_{3/2}$. J Chem Phys 141(24):244,310. https://doi.org/10.1063/1.4904265
56. Masnou-Seeuws F, Pillet P (2001) Formation of ultracold molecules (T $\leq$ 200 μK) via photoassociation in a gas of laser-cooled atoms. Adv At Mol Opt Phy 47:53–127. https://doi.org/10.1016/s1049-250x(01)80055-0
57. McGuyer BH, McDonald M, Iwata GZ, Tarallo MG, Grier AT, Apfelbeck F, Zelevinsky T (2015) High-precision spectroscopy of ultracold molecules in an optical lattice. New J Phys 17(5):055,004. https://doi.org/10.1088/1367-2630/17/5/055004
58. Meniailava DN, Shundalau MB (2017) Multi-reference perturbation theory study on the CsYb molecule including the spin-orbit coupling. Comput Theor Chem 1111:20–26. https://doi.org/10.1016/j.comptc.2017.03.046
59. Mickelson PG, Martinez YN, Saenz AD, Nagel SB, Chen YC, Killian TC, Pellegrini P, Côté R (2005) Spectroscopic determination of the s-wave scattering lengths of ^{86}Sr and ^{88}Sr. Phys Rev Lett 95(22). https://doi.org/10.1103/physrevlett.95.223002
60. Mulliken RS (1930) The interpretation of band spectra. Parts I, IIa, IIb. Rev Mod Phys 2(1):60–115. https://doi.org/10.1103/revmodphys.2.60
61. Münchow F, Bruni C, Madalinski M, Gorlitz A (2011) Two-photon photoassociation spectroscopy of heteronuclear YbRb. Phys Chem Chem Phys 13(42):18,734. https://doi.org/10.1039/c1cp21219b
62. Nemitz N, Baumer F, Münchow F, Tassy S, Görlitz A (2009) Production of heteronuclear molecules in an electronically excited state by photoassociation in a mixture of ultracold Yb and Rb. Phys Rev A 79(6):061,403. https://doi.org/10.1103/PhysRevA.79.061403

63. Pellegrini P, Gacesa M, Côté R (2008) Giant formation rates of ultracold molecules via Feshbach-optimized photoassociation. Phys Rev Lett 101(5):053,201. https://doi.org/10.1103/physrevlett.101.053201
64. Pichler M, Chen H, Stwalley WC (2004) Photoassociation spectroscopy of ultracold Cs below the $6P_{1/2}$ limit. J Chem Phys 121(4):1796–1801. https://doi.org/10.1063/1.1767071
65. Pichler M, Chen H, Stwalley WC (2004) Photoassociation spectroscopy of ultracold Cs below the $6P_{3/2}$ limit. J Chem Phys 121(14):6779–6784. https://doi.org/10.1063/1.1788657
66. Pichler M, Stwalley WC, Dulieu O (2006) Perturbation effects in photoassociation spectra of ultracold Cs_2. J Phys B: At, Mol Opt Phys 39(19):S981. https://doi.org/10.1088/0953-4075/39/19/S12
67. Pruvost L, Jelassi H (2010) Weakly bound $(6S_{1/2}+6P_{1/2})$ 0_g^- Cs_2 levels analysed using the vibrational quantum defect: detection of two deeply bound $(6S_{1/2}+6P_{1/2})$ 0_g^- levels. J Phys B: At, Mol Opt Phys 43(12):125,301. https://doi.org/10.1088/0953-4075/43/12/125301
68. Roy R, Shrestha R, Green A, Gupta S, Li M, Kotochigova S, Petrov A, Yuen CH (2016) Photoassociative production of ultracold heteronuclear YbLi* molecules. Phys Rev A 94(3):033,413. https://doi.org/10.1103/physreva.94.033413
69. Safronova MS, Budker D, DeMille D, Kimball DFJ, Derevianko A, Clark CW (2018) Search for new physics with atoms and molecules. Rev Mod Phys 90(025):008. https://doi.org/10.1103/RevModPhys.90.025008
70. Sage JM, Sainis S, Bergeman T, DeMille D (2005) Optical production of ultracold polar molecules. Phys Rev Lett 94(20):203,001. https://doi.org/10.1103/physrevlett.94.203001
71. Sivia D, Skilling J (2006) Data analysis: a Bayesian tutorial. Oxford University Press
72. Steck DA (2010) Cesium D line data. http://steck.us/alkalidata (revision 2.1.4)
73. Stellmer S, Pasquiou B, Grimm R, Schreck F (2012) Creation of ultracold Sr_2 molecules in the electronic ground state. Phys Rev Lett 109(11):115,302. https://doi.org/10.1103/PhysRevLett.109.115302
74. Sugawa S, Yamazaki R, Taie S, Takahashi Y (2011) Bose-Einstein condensate in gases of rare atomic species. Phys Rev A 84(1):011,610. https://doi.org/10.1103/PhysRevA.84.011610
75. Taie S, Takasu Y, Sugawa S, Yamazaki R, Tsujimoto T, Murakami R, Takahashi Y (2010) Realization of a $SU(2) \times SU(6)$ system of fermions in a cold atomic gas. Phys Rev Lett 105(19):190,401. https://doi.org/10.1103/PhysRevLett.105.190401
76. Takasu Y, Takahashi Y (2009) Quantum degenerate gases of ytterbium atoms. J Phys Soc Jpn 78(1):012,001. https://doi.org/10.1143/JPSJ.78.012001
77. Takasu Y, Maki K, Komori K, Takano T, Honda K, Kumakura M, Yabuzaki T, Takahashi Y (2003) Spin-singlet Bose-Einstein condensation of two-electron atoms. Phys Rev Lett 91(4):040,404. https://doi.org/10.1103/PhysRevLett.91.040404
78. Takasu Y, Komori K, Honda K, Kumakura M, Yabuzaki T, Takahashi Y (2004) Photoassociation spectroscopy of laser-cooled ytterbium atoms. Phys Rev Lett 93(123):202. https://doi.org/10.1103/PhysRevLett.93.123202
79. Thorpe JI, Numata K, Livas J (2008) Laser frequency stabilization and control through offset sideband locking to optical cavities. Opt Express 16(20):15,980. https://doi.org/10.1364/oe.16.015980
80. Tiesinga E, Williams C, Julienne P, Jones K, Lett P, Phillips W (1996) A spectroscopic determination of scattering lengths for sodium atom collisions. J Res Nat Inst Stand Technol 101(4):505. https://doi.org/10.6028/jres.101.051
81. Tolra BL, Drag C, Pillet P (2001) Observation of cold state-selected cesium molecules formed by stimulated Raman photoassociation. Phys Rev A 64(6):061,401. https://doi.org/10.1103/physreva.64.061401
82. Tolra BL, Hoang N, T'Jampens B, Vanhaecke N, Drag C, Crubellier A, Comparat D, Pillet P (2003) Controlling the formation of cold molecules via a Feshbach resonance. Europhys Lett 64(2):171–177. https://doi.org/10.1209/epl/i2003-00284-x
83. Udem T, Reichert J, Holzwarth R, Hänsch TW (1999) Absolute optical frequency measurement of the cesium d_1 line with a mode-locked laser. Phys Rev Lett 82(18):3568–3571. https://doi.org/10.1103/physrevlett.82.3568

84. Vanhaecke N, de Souza Melo W, Tolra BL, Comparat D, Pillet P (2002) Accumulation of cold cesium molecules via photoassociation in a mixed atomic and molecular trap. Phys Rev Lett 89(6):063,001. https://doi.org/10.1103/physrevlett.89.063001
85. Vanhaecke N, Lisdat C, T'Jampens B, Comparat D, Crubellier A, Pillet P (2004) Accurate asymptotic ground state potential curves of Cs_2 from two-colour photoassociation. Eur Phys J D 28(3):351–360. https://doi.org/10.1140/epjd/e2004-00001-y
86. Weiner J, Bagnato VS, Zilio S, Julienne PS (1999) Experiments and theory in cold and ultracold collisions. Rev Mod Phys 71(1):1–85. https://doi.org/10.1103/revmodphys.71.1
87. Wu J, Ma J, Zhang Y, Li Y, Wang L, Zhao Y, Chen G, Xiao L, Jia S (2011) High sensitive trap loss spectroscopic detection of the lowest vibrational levels of ultracold molecules. Phys Chem Chem Phys 13(42):18,921. https://doi.org/10.1039/c1cp22314c
88. Zabawa P, Wakim A, Haruza M, Bigelow NP (2011) Formation of ultracold $X^1\Sigma^+(v'=0)$ NaCs molecules via coupled photoassociation channels. Phys Rev A 84(061):401. https://doi.org/10.1103/PhysRevA.84.061401
89. Zelevinsky T, Kotochigova S, Ye J (2008) Precision test of mass-ratio variations with lattice-confined ultracold molecules. Phys Rev Lett 100(4):043,201. https://doi.org/10.1103/physrevlett.100.043201
90. Zuchowski PS, Hutson JM (2010) Reactions of ultracold alkali-metal dimers. Phys Rev A 81(6):060,703. https://doi.org/10.1103/PhysRevA.81.060703

Chapter 8
Two-Photon Photoassociation

With the one-photon photoassociation measurements complete, we now expand our binding energy measurements to the ground state. An accurate model of the electronic ground state potential is essential for accurate prediction of interspecies Feshbach resonances and can also be used to calculate interspecies scattering lengths. Ab initio calculations of these molecular potentials are challenging and cannot achieve the accuracy required for our purposes. Therefore, to better constrain a model of the CsYb molecular potential we require precise measurements of binding energies of near-threshold vibrational levels of the electronic ground state.

Two-photon photoassociation is an extremely useful technique to measure the binding energy of near-threshold vibrational levels. This technique has been applied in a large number of single-species [1, 2, 12, 17, 22, 30, 34, 44, 46, 48] and two-species ultracold atom experiments [10, 11, 18, 31, 32, 38] with considerable success. In addition, pairing of this technique with Autler-Townes spectroscopy allows measurements of the FCFs between electronically excited and ground state vibrational levels. Knowledge of FCFs is essential for identification of appropriate transitions for coherent transfer to the ground state.

In this chapter we perform two-photon photoassociation spectroscopy to measure the binding energies of vibrational levels of the CsYb $X\,^2\Sigma_{1/2}^+$ ground state. The measurements are performed on three isotopologs $^{133}Cs^{170}Yb$, $^{133}Cs^{173}Yb$ and $^{133}Cs^{174}Yb$, in each case we determine the energy of several vibrational levels including the least-bound state. An interaction potential is fitted to the measurements that allows the prediction of interspecies scattering lengths for all seven CsYb isotopologs. The results presented in this chapter are published in Ref. [19].

8.1 Two-Photon Photoassociation of CsYb

The two-photon photoassociation process is depicted in Fig. 8.1. This scheme is an extension of the one-photon photoassociation scheme we considered in the previous chapter. The laser that drives the one-photon photoassociation, L_1, has frequency

© Springer Nature Switzerland AG 2019
A. Guttridge, *Photoassociation of Ultracold CsYb Molecules
and Determination of Interspecies Scattering Lengths*, Springer Theses,
https://doi.org/10.1007/978-3-030-21201-8_8

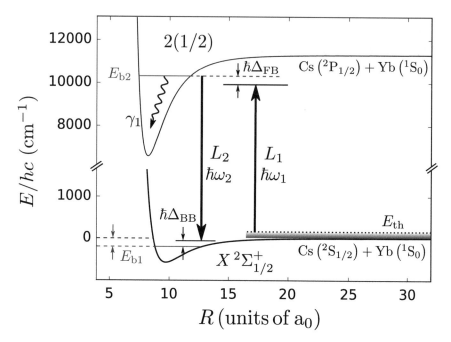

Fig. 8.1 Two-photon photoassociation for the measurement of the binding energy, E_{b1}, of a vibrational level of CsYb in its electronic ground state. A pair of colliding Cs and Yb atoms with thermal energy E_{th} is associated to form a CsYb molecule in a rovibrational level of the electronically excited 2(1/2) state by light of frequency ω_1. This rovibrational level is coupled to a level in the electronic ground state, $X\,^2\Sigma_{1/2}^+$, by light of frequency ω_2. The molecular curves plotted here are adapted from Ref. [29]. The internuclear distances where the transitions occur are not depicted to scale

ω_1 and is detuned from a free-bound transition by Δ_{FB}. The second laser, L_2, has frequency ω_2 and couples the electronically excited molecule to a rovibrational level of the electronic ground state. The detuning from this bound-bound transition is labelled as Δ_{BB}. When the laser L_2 is resonant with a bound-bound transition, the coupling leads to the formation of a dark state and the suppression of the absorption of L_1. Such two-photon dark-resonances can be used to measure the binding energies, E_{b1}, of vibrational levels of the molecule in the electronic ground state. In the undressed, zero-temperature limit, the binding energy is given simply by the difference in photon energy of the two lasers on two-photon resonance

$$E_{b1} = \hbar\,(\omega_2 - \omega_1). \qquad (8.1)$$

For the specific case of CsYb, the first photon excites the colliding atoms into a rovibrational level of the molecule close to the $Cs(^2P_{1/2}) + Yb(^1S_0)$ asymptote. We use the same electronic state, 2(1/2), as in the one-photon photoassociation measurements. A second photon couples the vibrational level of the electronically excited state to a near-threshold rovibrational level of the $X\,^2\Sigma_{1/2}^+$ ground state. A combination of selection rules and the low temperature of our Cs and Yb atomic mixture, result in all the rovibrational levels we measure having $\ell = 0$. Therefore, we label each molecular level by its vibrational number n below the associated threshold, using n' for the electronically excited state and n'' for the ground state, such that $n = -1$ corresponds to the least-bound state.

8.1.1 Experimental Setup

The experimental conditions and hardware are essentially unchanged from the one-photon photoassociation measurements. The major difference in the two-photon case is the need for two lasers, one to drive the free-bound transition and the other to drive the bound-bound transition. The Ti:Sapphire laser used in the one-photon measurements is primarily used to drive the bound-bound transition as it is tunable over a greater frequency range. We use the DBR laser (used for the Cs MOT measurements in Sect. 7.3.1) to drive the free-bound transitions.

A schematic illustrating the setup of both lasers is shown in Fig. 8.2. In this locking scheme, both lasers are stabilised to the same optical cavity. The light from L_1 and L_2 pass through independent fibre EOMs which add frequency sidebands to the light. After the output of the fibre EOMs, light from L_1 and L_2 is combined on a 50 : 50 beam splitter and the overlapped beams are coupled down the same optical fibre to the optical cavity. The electronic sideband technique is used to allow continuous tuning of both laser frequencies by stabilising a frequency sideband to a cavity transmission

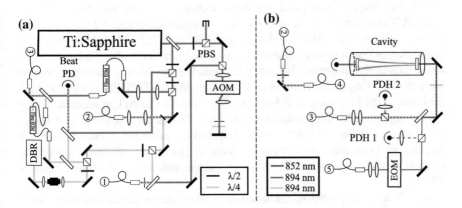

Fig. 8.2 Two-photon photoassociation laser setup. **a** Setup of the Ti:Sapphire and DBR lasers on the PA optical table. Fibre 1 carries light to the experiment. **b** Stabilisation of photoassociation light to high finesse cavity on the laser table. Fibre 5 is the light from the Cs ZS AOM used to stabilise the cavity length. Fibre 4 carries light to the wavemeter

peak. By stabilising the lasers to different modes of the cavity we can control their frequency difference, $\omega_1 - \omega_2$, over a large dynamic range.

The main output of the Ti:Sapphire laser is double passed through an AOM in the same setup as the previous chapter. The main output of the DBR laser is over-lapped with the Ti:Sapphire light using a beamsplitter and both beams pass through a shutter before they are coupled into a fibre which carries the light to the exper-iment. The output of the fibre is focused onto the trapped atomic mixture with a waist of 150 μm and is circularly polarised to drive σ^+ transitions. As we saw in Sect. 7.4.2, σ^+ polarisation yields the strongest two-photon transitions from the $Cs(6^2 S_{1/2} F = 3, m_F = +3) + Yb(^1 S_0)$ scattering state to the $F = 3$ manifold of the $X^2 \Sigma_{1/2}^+$ molecular ground state via an intermediate vibrational level of CsYb in the $F' = 4$ manifold of the $2(1/2)$ potential.

Using two-photon photoassociation, binding energies may be measured to sub-MHz precision. At this level of precision many smaller factors can influence the the measurement, one of these is the Doppler shifts of the ultracold atomic mixture. For our initial temperatures, Doppler shifts of order $\sim$200 kHz may occur. The Doppler shift of a two-photon transition is

$$\Delta_{Doppler} = (\Delta'_{BB} - \Delta'_{FB}) - (\Delta_{BB} - \Delta_{FB})$$
$$= (\Delta'_{BB} - \Delta_{BB}) - (\Delta'_{FB} - \Delta_{FB}) = (\mathbf{k}_{BB} - \mathbf{k}_{FB}) \cdot \mathbf{v}. \qquad (8.2)$$

It is important to note that the Doppler shift is therefore minimised for the co-propagating geometry used in these measurements.

We measure the frequency difference between lasers L_1 and L_2 using one of three methods, depending on the binding energy of the state under investigation. Most generally, the frequency difference is determined from the difference in the modulation frequencies applied to the two EOMs, combined with the number of cavity FSRs between the two modes used for frequency stabilisation. Light from both lasers is coupled into a commercial wavemeter (Bristol 671A) for absolute frequency calibration and unambiguous determination of the cavity mode. For binding energies below 2 GHz, the frequency difference between the two lasers is measured directly from the beat frequency recorded on a fast photodiode (EOT ET-2030A). In both these cases, the frequency shift from the AOM must be taken into account as only the Ti:Sapphire light passes through the AOM. In the special case of the least-bound state, we do not use the DBR laser and instead we drive the AOM with two RF frequencies. Generating the two-photon detuning in this way eliminates any effects of laser frequency noise and allows a very precise determination of the frequency difference.

At this level of precision, the home-built AOM driver circuits based on VCO chips may effect the measurements of the binding energies. We measure the driving frequency of the AOM directly on a frequency counter and find the frequency of the VCO is stable to $\sim$kHz over the course of a day. However, we do notice shifts of order $\sim$10s of kHz when changing the amplitude of the sine wave generated by the VCO. Therefore, when taking measurements of near-threshold levels it is important to measure the AOM frequency for each power applied to the AOM.

8.1.2 Spectroscopy Methods

The molecular states represented in Fig. 8.1 are analogous to the three-level lambda system previously considered during our discussion of STIRAP in Sect. 2.5. The quantum interference effects that arise in three-level lambda systems have been extensively studied in atomic physics [14] and these effects have been exploited in molecular systems to allow detection of molecular levels through dark state spectroscopy [10, 28, 30, 50] and efficient transfer of molecules to the ground state using STIRAP [3, 47]. The creation of dark states is essential to the spectroscopy we consider here so we repeat the results derived in Eq. 2.9:

$$|a^+\rangle = \sin\theta\sin\phi|E_{th}\rangle + \cos\phi|E_{b2}\rangle + \cos\theta\sin\phi|E_{b1}\rangle, \tag{8.3a}$$

$$|a^0\rangle = \cos\theta|E_{th}\rangle - \sin\theta|E_{b1}\rangle, \tag{8.3b}$$

$$|a^-\rangle = \sin\theta\cos\phi|E_{th}\rangle - \sin\phi|E_{b2}\rangle + \cos\theta\cos\phi|E_{b1}\rangle. \tag{8.3c}$$

Here the two mixing angles θ and ϕ are defined by

$$\tan\theta = \frac{\Omega_{FB}}{\Omega_{BB}}, \qquad \tan 2\phi = \frac{\sqrt{\Omega_{FB}^2 + \Omega_{BB}^2}}{\Delta_{2\gamma}}, \tag{8.4}$$

where, we have defined the two photon detuning $\Delta_{2\gamma} = \Delta_{FB} - \Delta_{BB}$.

To obtain measurements of the ground state binding energies we employed two distinct techniques. Here we briefly summarise both techniques.

Dark-Resonance Spectroscopy

In dark-resonance spectroscopy the elimination of the absorption of a probe laser, in our case L_1, is detected when a second coupling laser, L_2, is resonant with a bound-bound transition. In this scheme we require $\Omega_{BB} \gg \Omega_{FB}$, which is readily achieved with modest laser powers due to the small Franck-Condon overlap of the free-bound transition. In our molecular system, the absorption of L_1 is monitored using the detected number of Cs atoms. When L_1 is resonant with a free-bound transition, Cs*Yb molecules are produced resulting in a drop in the number of Cs atoms detected by absorption imaging. As L_2 is tuned into resonance with the bound-bound transition, we have the condition $\sin\theta \to 0$ and $\cos\theta \to 1$. This results in $|a^0\rangle = |E_{th}\rangle$, so the atomic scattering state is no longer coupled to the excited molecular state. Therefore, the production of Cs*Yb molecules is interrupted and we regain our Cs atom number. The application of dark-resonance spectroscopy to the measurement of the least-bound state of the ^{133}Cs^{174}Yb molecule in the $X\,^2\Sigma_{1/2}^+$ ground state is shown in Fig. 8.3.

Figure 8.3a shows the lineshape observed when the frequency of L_2 is fixed on resonance with the bound-bound transition ($\Delta_{BB} = 0$) and the frequency of L_1 is scanned over the free-bound transition. The spectrum exhibits the w-shaped profile expected for electromagnetically induced transparency (EIT) in a lambda-type three-

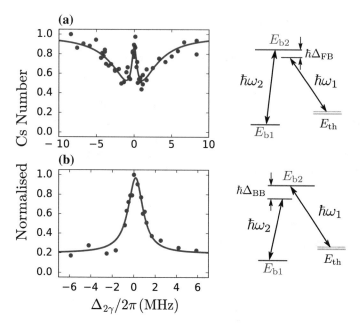

Fig. 8.3 Measurement of the least-bound state of the $^{133}\mathrm{Cs}^{174}\mathrm{Yb}$ molecule in the $X\,^2\Sigma^+_{1/2}$ ground state by dark-resonance spectroscopy. Left panel: Two-photon photoassociation spectra. Normalised number of Cs atoms plotted against $\Delta_{2\gamma} = \Delta_{\mathrm{FB}} - \Delta_{\mathrm{BB}}$. Right Panel: Simplified level structure for the two-photon photoassociation transitions. **a** Dark-resonance spectroscopy performed by scanning the frequency of L_1. The red solid line shows a fit to the data using Eq. 8.5. **b** Dark-resonance spectroscopy performed by scanning the frequency of L_2. The red solid line is a fit using a Lorentzian profile. The spectra shown in (**a**) and (**b**) were obtained with laser intensities $I_1 = 0.42$ and $0.68\,\mathrm{W/cm^2}$ and $I_2 = 0.79$ and $0.57\,\mathrm{W/cm^2}$ respectively

level system [14] and we therefore refer to this as the EIT lineshape. When ω_1 is off-resonant, the two photon detuning is no longer zero and a fraction of E_{b2} is mixed into the dark state. In the wings of the observed lineshape we identify a Lorentzian profile originating from one-photon photoassociation to the $n' = -13$ level of the molecule in the 2(1/2) potential. Then, on resonance we see a suppression of the photoassociative loss due to the creation of an almost-pure dark state composed of the initial atomic scattering state and a small admixture of the molecular ground state. This dark state is decoupled from the intermediate $n' = -13$ state and leads to the observed 'transparency'. The Hamiltonian derived in Sect. 2.5 may be used to model this system. However, it is simpler to use an analytical solution of the optical Bloch equations for a lambda-type three-level system [10, 14] in the limit of $\Omega_{\mathrm{BB}} \gg \Omega_{\mathrm{FB}}$,

$$\frac{N}{N_0} = \exp\left\{-\frac{t_{\mathrm{PA}}\,\Omega^2_{\mathrm{FB}}(4\Gamma\Delta^2_{2\gamma} + \Gamma_{\mathrm{eff}}(\Omega^2_{\mathrm{BB}} + \Gamma_{\mathrm{eff}}\Gamma))}{|\Omega^2_{\mathrm{BB}} + (\Gamma + 2i\,\Delta_{\mathrm{FB}})(\Gamma_{\mathrm{eff}} + 2i\,\Delta_{2\gamma})|^2}\right\}. \qquad (8.5)$$

Here, t_{PA} is the irradiation time of the photoassociation lasers, Ω_{FB} (Ω_{BB}) is the Rabi frequency on the free-bound (bound-bound) transition and Γ_{eff} is a phenomenological constant that accounts for the decoherence of the dark state. From the fit of this equation to the data we find $\Omega_{BB}/2\pi = 2.0(5)\,\mathrm{MHz}$.

Figure 8.3b shows the dark-resonance spectrum observed when the frequency of L_1 is resonant with the free-bound transition ($\Delta_{FB} = 0$) and the frequency of laser L_2 is scanned. This case complements the EIT lineshape shown in Fig. 8.3a, where the only difference between the two cases is which laser's frequency is scanned.

Off resonance with the bound-bound transition we observe a large loss of Cs atoms due to the production of Cs*Yb molecules. When L_2 is tuned close to resonance with the bound-bound transition we regain our dark state where L_1 is no longer resonant with the free-bound transition. Therefore, the production of Cs*Yb molecules is suppressed and there is a recovery in the Cs number. By setting $\Delta_{FB} = 0$ in Eq. 8.5 and neglecting Γ_{eff} in the limit $\gamma\Gamma_{eff} \ll \Omega_{BB}^2$ we obtain

$$\frac{N}{N_0} = \exp\left(-t_{PA}\frac{\Omega_{FB}^2}{\gamma}\frac{4\gamma\Delta_{2\gamma}^2}{\Omega_{BB}^4 + 4\gamma^2\Delta_{2\gamma}^2}\right). \tag{8.6}$$

When the number of atoms lost is small, $t_{PA}\Omega_{BB}^2/\gamma \ll 1$, the exponent is small and $\exp(-x) \approx 1 - x$. In this case a Lorentzian profile is obtained and the width of the profile is determined solely by Ω_{BB}.

This dark-resonance technique is the simplest method for the observation of a two-photon resonance, as with sufficient L_2 intensity the feature can be significantly broadened. By using the Ti:Sapphire laser to drive the bound-bound transition we can significantly broaden the transition and more quickly search a large frequency range. However, the background number of Cs atoms is sensitive to the one-photon photoassociation loss rate and can therefore drift in response to changes in the Yb density, the Cs density, or the photoassociation light intensity or polarisation.

Raman Spectroscopy

Another method for observing a two-photon resonance is to perform Raman spectroscopy. In this case, the frequency of a single PA laser is fixed at a certain detuning from a free-bound or bound-bound transition and the frequency of the other PA laser is scanned. A Raman transition is driven to the ground-state vibrational level when the Raman condition is fulfilled ($\Delta_{FB} = \Delta_{BB}$). This two-photon process can proceed only by a virtual excitation in the first step and results in a narrow lineshape.

In this regime, the molecular energy levels are only weakly perturbed by the laser fields in comparison to the dark-resonance spectroscopy method. The expected lineshape for cold atom scattering in the presence of two laser fields, with one of the lasers off-resonance, is shown in Fig. 8.4. In this case L_2 is detuned from the bound-bound transition and ω_1 is scanned. In the presence of only a single laser field (blue dashed) we observe the expected Lorentzian feature corresponding to one-photon photoassociation. However, when the coupling of L_2 is increased to $\Omega_{BB}/2\pi = 5\,\mathrm{MHz}$, we observe a shift of the one-photon line due to an AC Stark shift

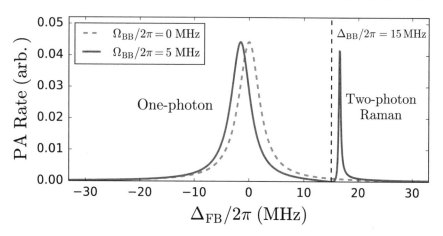

Fig. 8.4 Photoassociation rate as a function of detuning from a one-photon photoassociation transition Δ_{FB} for a bound-bound Rabi frequency $\Omega_{BB}/2\pi = 0$ MHz (blue dashed) and $\Omega_{BB}/2\pi = 5$ MHz (red solid). L_2 is detuned from the bound-bound transition by $\Delta_{BB}/2\pi = 15$ MHz, this detuning is shown by the dashed black line

induced by L_2. We also observe the appearance of a narrow feature corresponding to a two-photon Raman transition. In Raman spectroscopy we measure the binding energy of the ground state vibrational level using the loss maximum. The loss feature occurs when the relative laser energy $\hbar(\omega_1 - \omega_2)$ is equal to the energy difference between the light-shifted states $E'_{b2} - E'_{th}$.

Interestingly, the Raman resonance also exhibits a dark-resonance effect when the Raman condition is fulfilled between the unperturbed states. The black dashed line shows the detuning of the L_2 from the unperturbed bound-bound transition, where a small suppression of loss is evident. Like the features discussed earlier, this dark-resonance feature is unaffected by light shifts from the PA lasers. However, use of this Raman dark-resonance to measure the binding energy is not practical. The loss suppression occurs at a point where the background loss is small, so excellent signal-to-noise is required to observe this feature. A more practical method of measuring the binding energy is by fitting to the loss maximum and accounting for the systematic light shifts due to the PA lasers.

Figure 8.5 shows the observation of a CsYb two-photon resonance using Raman spectroscopy. In contrast to Fig. 8.4, the frequency of L_1 is detuned from the free-bound transition ($\Delta_{FB} = -15$ MHz) and ω_2 is scanned. In this regime we are in the wings of the one-photon lineshape and we only observe the Raman feature as we are scanning ω_2. A loss of Cs atoms is observed when the lasers drive a two-photon transition to the electronic ground state. This corresponds to the creation of a ground-state CsYb molecule which is dark to our imaging and causes the observed decrease in the number of Cs atoms.

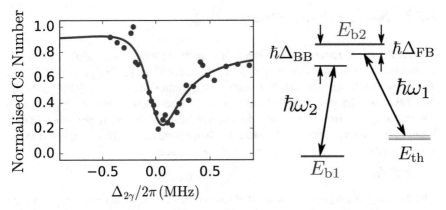

Fig. 8.5 Measurement of the least-bound state of the $^{133}Cs^{174}Yb$ molecule in the $X\,^2\Sigma^+_{1/2}$ ground state by Raman spectroscopy. Left Panel: Normalised number of Cs atoms as a function of two-photon detuning $\Delta_{2\gamma} = \Delta_{FB} - \Delta_{BB}$. The spectrum was obtained with laser intensities $I_1 = 1.1\,W/cm^2$ and $I_2 = 0.76\,W/cm^2$. The red solid line shows a fit to the data using the Fano profile. Right Panel: Simplified level structure for the Raman transition

Using the Raman spectroscopy technique we extract the line centre from the loss feature by fitting a Fano profile [13] to the data. A Fano lineshape arises in this system from the interference between two paths ($E_{th} \rightarrow E_{b_2}$ and $E_{th} \rightarrow E_{b_2} \rightarrow E_{b_1} \rightarrow E_{b_2}$) coupling the initial state to the continuum of decay products associated with E_{b_2} [4, 35].[1] Here, the second path passes through an intermediate discrete state (E_{b_1}) leading to the Fano profile.

It is still reasonable to wonder if the asymmetry in the observed Raman feature is in fact due to the distribution of energies in the initial atomic collision, an effect which is observed in photoassociation measurements at higher temperatures [21]. We have verified that the width of the feature is a result of coupling from the laser fields by observing the width as a function of L_2 intensity. For low intensities the feature depth is low and fitting is more difficult but we observe a width of $\sim$200 kHz. We surmise for lower powers the spectra will be dominated by thermal broadening, with a width of $\sim$70 kHz expected given our experimentally determined temperatures.

The interference effects observed in the two-photon spectra implies we have achieved coherent coupling between atomic and molecular states, however, we have not been able to observe Rabi oscillations in this experimental configuration. This is consistent with observations of other two-photon photoassociation experiments using electric-dipole transitions [37], where off-resonant excitation of atoms and molecules leads to rapid dephasing. In this regard, exploration of transitions to molecular states in the spin-forbidden Yb triplet manifold may be useful for coherent production of ground state CsYb molecules. Rabi oscillations between atomic and molecular states have been observed in homonuclear one-photon PA below the Yb $^3P_2$ atomic threshold [40].

---

[1]Rather confusingly, the continuum is not that of the initial collision energies but of E_{b_2} due to its much larger energy width.

8.1.3   Binding Energy Measurements

We use Raman spectroscopy as the primary method for the observation of $n'' = -1$ levels, as the lineshape of the two-photon feature is narrow for low powers of L_1 and L_2. The width of the observed resonance is proportional to $\Gamma_{\mathrm{Raman}} = (2h\Omega_{\mathrm{BB}}/\Delta_{\mathrm{BB}})^2$ [4]. Therefore, larger detuning from resonance leads to weaker coupling and narrower linewidths. However, a large detuning reduces the feature depth which in turn impacts on the uncertainty in the extracted line centre. The powers employed in the spectroscopy are a trade-off between these two effects.

Light Shifts

The coupling of the ground and excited states by L_1 and L_2 cause the energies of the states to shift. In the case of the dark-resonance lineshape this shift is the origin of the observed transparency on two-photon resonance. Therefore, measurements of the binding energy using the dark-resonance spectroscopy method are unaffected by light shifts so long as external levels are not introduced into the three-level system [51].

Conversely, in measurements using Raman transitions, the Raman peak (maximum Cs loss) is perturbed by light shifts. The line shift of the transition is a linear function of laser intensity in the perturbative limit [35]. Figure 8.6 shows the shifts $\delta_1(I_1)$ and $\delta_2(I_2)$ of the two-photon resonance position as functions of the intensity of lasers L_1 and L_2. We fit a straight line to the data to extract the line position at zero intensity, corresponding to the energy difference of the unperturbed states. As expected, the gradient of the shift with intensity is largest for the bound-bound transition, due to the much larger FCF between two bound levels compared to the FCF between the scattering state and an excited vibrational level.

Other Systematic Effects

Further systematic effects that may shift the position of the Raman line are: the AC Stark shift due to the dipole trapping light, the Zeeman effect of the magnetic field and the finite energy of the initial atomic collision.

The trapping light may systematically shift the line position by a differential AC Stark shift between the atomic pair and the molecular state E_{b1}. However, this shift is expected to be small for the weakly bound states considered here as the molecular polarisability is near identical to the atomic polarisability for near-threshold levels. The effect of magnetic field on the results is also expected to be small, as the linear Zeeman shift is almost the same for the atomic state and the molecular state. Investigation of shifts due to both magnetic field and dipole trap intensity found no significant shift at the resolution of the measurements (<100 kHz).

The remaining systematic shift in our measurements is the thermal shift, E_{th}, due to the energy of the initial collision between the Cs and Yb atom. We account for this effect by subtracting the mean collision energy from our binding energy measurements. The mean collision energy of the Cs+Yb atom pair is

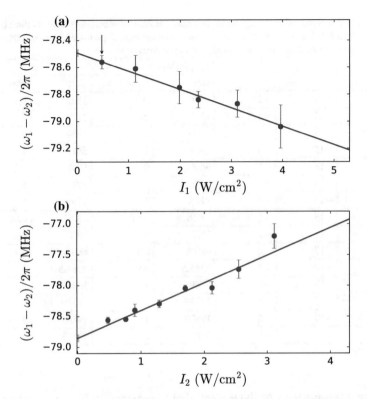

Fig. 8.6 Light shift of the $Cs^{174}Yb$ $n'' = -1$ Raman line as a function of photoassociation laser intensity, using the $n' = -13$ intermediate state. **a** Measured line center frequency as a function of intensity I_1 of laser L_1 driving the free-bound transition. The intensity of laser L_2 for this data set was $I_2 = 0.35 \, \mathrm{W/cm^2}$. **b** Measured line center frequency as a function of intensity I_2 of laser L_2 driving the bound-bound transition. The intensity of laser L_1 for this data set was $I_1 = 0.48 \, \mathrm{W/cm^2}$ and is highlighted in (**a**). The 1σ uncertainties in the intercepts are represented by the shaded regions at the origins

$$E_{\mathrm{th}} = \frac{3}{2}\mu_r k_B \left(T_{\mathrm{Yb}}/m_{\mathrm{Yb}} + T_{\mathrm{Cs}}/m_{\mathrm{Cs}}\right) \qquad (8.7)$$

where μ_r is the reduced mass. For our initial temperatures of $T_{\mathrm{Yb}} = 1 \, \mu\mathrm{K}$ and $T_{\mathrm{Cs}} = 6 \, \mu\mathrm{K}$, the correction is of order $100 \, \mathrm{kHz}$ and so only significantly affects the measurements of the $n'' = -1$ levels. A different thermal shift correction is applied to each isotopolog due to the small difference in Cs temperature during the measurements.

Binding Energies

In total we observed 14 ground-state vibrational levels using three isotopologs $^{133}Cs^{170}Yb$, $^{133}Cs^{173}Yb$ and $^{133}Cs^{174}Yb$. The binding energies of these levels, corrected for thermal and light shifts, are listed in Table 8.1. Both Raman and dark-

Table 8.1 Observed binding energies and their uncertainties for vibrational levels of three different isotopologs of CsYb in its electronic ground state, together with experimental 1σ uncertainties and binding energies calculated from the fitted interaction potential

Yb Isotope	n'	n''	E_{b1}/h (MHz)			
			Obs	Uncertainty	Calc	Obs$-$Calc
170	-15	-1	15.7	0.3	15.6	0.1
170	-15	-3	1576	2	1576	0
170	-15	-4	4259	2	4257	2
170	-15	-5	8988	2	8989	1
173	-13	-1	56.8	0.2	57.0	0.2
173	-13	-2	592	1	591	1
173	-13	-3	2166	1	2165	1
174	-13	-1	78.66	0.09	78.73	0.07
174	-17	-1	78.7	0.1	78.7	0.0
174	-17	-2	686.4	0.7	686.5	0.1
174	-17	-3	2385.5	0.9	2384.5	1
174	-17	-4	5749	1	5747	2
174	-17	-5	11358	1	11359	1
174	-17	-6	19803	1	19805	2
174	-17	-7	31672	2	31668	4

resonance spectroscopy methods were used to measure the binding energies of all ^{133}Cs^{174}Yb levels. The dark-resonance spectroscopy method where the frequency of laser L_2 is scanned was used for measurements of the $n'' < -1$ levels of other isotopologs for its simplicity, as only a single high resolution scan is necessary to find the binding energy of the vibrational level. Raman spectroscopy of $n'' < -1$ levels is not as precise due to broadening of the line by the relative frequency jitter of the Ti:Sapphire and DBR lasers.

The smaller error bars for the $n'' = -1$ levels result from the narrower Raman feature and the different method of generating the small frequency offset between the two photons. Laser frequency instabilities due to beating between the sidebands of L_1 and L_2 prevented observation of the $n'' = -2$ state of Cs^{170}Yb. The $n'' = -1$ state of ^{133}Cs^{174}Yb was measured with both $n' = -13$ and $n' = -17$ as intermediate states to verify that the measurements are of the ground electronic state and not a two-photon transitions to a higher-energy electronic state. We chose to use intermediate states with moderately large binding energy to increase the detuning of the photoassociation light from the Cs D_1 transition; a greater feature depth is observed for larger detuning due to the reduction of off-resonant Cs losses as discussed in Chap. 7.

8.2 Line Strengths and Autler-Townes Spectroscopy

The strengths of transitions between the electronically excited state and ground state may be determined from the light shift of the Raman spectroscopy measurements. The systematic dependencies of Raman transitions in three-level lambda-type systems have been studied extensively [4, 5, 27, 33, 51]. For atomic systems it has been shown that the light shift is proportional to Ω^2, where Ω is the Rabi frequency associated with either one-photon transition [5, 33]. Investigations of molecular systems have found that the light shift of the resonance maintains this Ω^2 dependence even in the presence of decay out of the three-level system [9, 35]. Here we wish to determine the line strengths for the bound-bound transitions given by Ω^2_{BB}/I_2 using the light-shift measurements of the type presented in Fig. 8.6b.

For the Raman lineshape shown in Fig. 8.5, the maximum loss of Cs atoms occurs at a two-photon detuning $\omega_1 - \omega_2 = E_{b1}/\hbar + \delta_1(I_1) + \delta_2(I_2)$. Here $\delta_1(I_1)$ and $\delta_2(I_2)$ are the light shifts of the transition and [35]

$$\frac{\delta_2(I_2)}{I_2} = \left(\frac{\Omega^2_{BB}/I_2}{4\Delta^2_{FB} + \Gamma^2} \right) \Delta_{FB}, \qquad (8.8)$$

where $\Delta_{FB} \simeq \Delta_{BB}$ in the vicinity of the Raman resonance.[2] It follows that the line strength Ω^2_{BB}/I_2 may be obtained from the gradient of resonance position with respect to intensity I_2 using Eq. 8.8. The results for the measured line strengths of $n' = -17 \rightarrow n''$ transitions in $^{133}Cs^{174}Yb$ are presented in Fig. 8.7 as green open circles.

The line strength of the bound-bound transitions may also be determined using Autler-Townes spectroscopy (ATS) to directly measure the Rabi frequency, Ω_{BB}, from the splitting of the two dressed states. The experimental configuration for ATS is the same as in Fig. 8.3a, but instead of measuring the binding energy we measure the splitting of the dressed states as a function of the intensity of L_2. Figure 8.8 shows the Autler-Townes spectrum of the $n' = -17 \rightarrow n'' = -5$ transition in $^{133}Cs^{174}Yb$. In the figure, ω_2 is fixed on resonance ($\Delta_{BB} = 0$) and ω_1 is scanned over the free-bound $n' = -17$ transition for a number of different intensities of L_2. The Autler-Townes splitting of the one-photon line is clearly visible as the intensity of the bound-bound laser is increased. The Rabi frequency is extracted by fitting Eq. 8.5 to the data, but may also be found by measuring the splitting of the two peaks (when they are well resolved). The quantity of interest, $\Omega_{BB}/\sqrt{I_2}$, is then extracted from a linear fit to the Rabi frequency as a function of square-root intensity as shown in Fig. 8.8b. We find that, for the $n' = -17 \rightarrow n'' = -5$ transition, $\Omega_{BB}/\sqrt{I_2} = 2\pi \times 19(1)\,\mathrm{MHz}/\sqrt{\mathrm{W\,cm^{-2}}}$. We include this measurement in Fig. 8.7 as the red closed circle. Note, we did not measure all the transitions using ATS due to

---

[2]The definition of Ω_{BB} in Eq. 8.8 is twice that in Ref. [35].

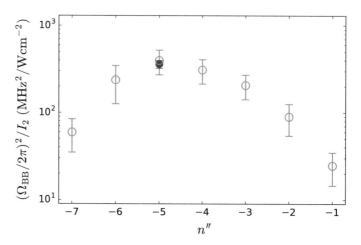

Fig. 8.7 Measured line strengths of $n' = -17 \rightarrow n''$ transitions in Cs^{174}Yb. The line strength Ω_{BB}^2/I_2 is plotted as a function of ground-state vibrational level n''. Green open circles represent measurements of the Rabi frequencies from the line shifts of the Raman loss features. The filled red circle represents the measurement of the $n' = -17 \rightarrow n'' = -5$ transition using Autler-Townes spectroscopy

the ~30 s load-detection cycle associated with conducting the measurements. Nevertheless, the excellent agreement between the two measurements of the line strength for the $n' = -17 \rightarrow n'' = -5$ transition confirms the validity of using the light-shift measurements.

The FCFs that determine the line strengths are dominated by the region around the outermost lobe of the wavefunction for $n' = -17$. This is far inside the outer turning points of the near-threshold levels of the ground electronic state. In this region, the wave functions of the different near-threshold levels in the electronic ground state are almost in phase with one another, but with amplitudes proportional to $E_{\text{bl}}^{1/3}$ [24] and line strengths proportional to $E_{\text{bl}}^{2/3}$. However, the wave functions start to change phase as E_{bl} increases; eventually the phase difference between the wave functions in the two electronic states overcomes the amplitude factor and the FCF starts to decrease. Figure 8.7 shows that the peak line strength occurs around $n'' = -5$ in the present case.

Previously, in Sect. 7.4.3, we fitted the one-photon photoassociation spectra to a near-dissociation expansion. However, the quantities C_6 and C_8 resulting from this are *effective* dispersion coefficients that incorporate higher-order effects. They are not sufficient to determine the outer turning point accurately at the energy of the $n' = -17$ level, which is bound by 286 GHz. Calculating FCFs will require a more complete model of the excited-state potential, which is beyond the scope of this investigation.

Fig. 8.8 Autler-Townes spectroscopy (ATS) of the $n' = -17 \rightarrow n'' = -5$ transition in $Cs^{174}Yb$. **a** Normalized Cs number versus detuning, Δ_{FB}, of laser L_1 from the $n' = -17$ free-bound transition. The second laser L_2 is on resonance with the bound-bound transition, $\Delta_{BB} = 0$, and the splitting of the one-photon lineshape is observed for varying intensities I_2 of laser L_2. **b** Bound-bound Rabi frequency Ω_{BB} extracted from the ATS measurements as a function of the square root of the intensity I_2 of laser L_2 that drives the bound-bound transition. The solid line is a linear fit with the intercept constrained to be zero

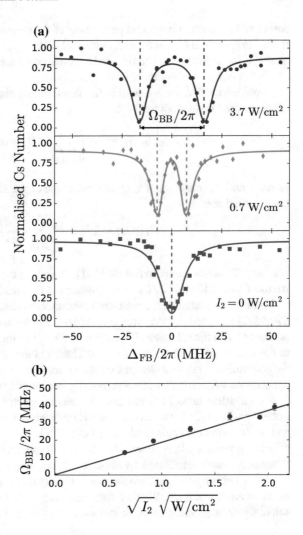

8.3 CsYb Ground State Molecular Potential

In modelling the one-photon photoassociation measurements in the previous chapter we used a long-range dispersive potential. However, this approach should not be used here as one of the main goals of the two-photon measurements is to guide our search for interspecies Feshbach resonances. Calculations of Feshbach resonance widths [6, 52] require a complete interaction potential, rather than just the long-range form. To obtain such a potential, the short-range part is based on electronic structure calculations. We decided to base our short-range potential on that of Brue and Hutson [6], this is the potential used in our previous fit to the thermalisation measurements in Sect. 6.4. However, the accuracy of the thermalisation measurements were not sufficient to constrain important parameters of the potential such as number of vibrational levels or the C_6 coefficient. A critical difference in the two-photon measurements is the ability to fit the molecular potential to the observed binding energies and better constrain our model of the potential.

In order to adjust the potential to fit our measured binding energies, the potential is represented in an analytic form,

$$V(R) = Ae^{-\beta R} - \sum_{n=6,8,10} D_n(\beta R) C_n R^{-n}. \tag{8.9}$$

Here, A and β control the magnitude and range of the short-range repulsive wall of the potential and

$$D_n(\beta R) = 1 - e^{-\beta R} \sum_{m=0}^{n} \frac{(\beta R)^m}{m!} \tag{8.10}$$

is a Tang-Toennies damping function [41]. To reduce the number of free parameters, we use $C_{10} = (49/40) C_8^2 / C_6$ as recommended by Thakkar and Smith [42].

To fit the potential to the measured binding energies, the dispersion coefficients C_6 and C_8 are fitted and A is varied to adjust the volume of the potential and thus the number of vibrational levels. We fix $\beta = 0.83 \, a_0^{-1}$ to the value obtained from fitting to the electronic structure calculations. These choices allow us to fit the aspects of the potential that are well-determined by our measurements, using a small number of parameters, while maintaining a physically reasonable form for the entire potential.

We calculate near-threshold bound states supported by the potential using the BOUND package [20]. The terms in the Hamiltonian that couple different electronic and nuclear spin channels (and cause Feshbach resonances) are very small [6]. The effective potential is thus almost identical for all spin channels. The bound molecular states are almost unaffected by these weak couplings. The effects of the atomic hyperfine splitting and Zeeman shifts are already accounted for in the measurement of the binding energies. We therefore calculate bound states using single-channel calculations, neglecting electron and nuclear spins and the effects of the magnetic field.

We carry out separate least-squares fits to the measured binding energies for each plausible number of vibrational levels N_{vib}. For the isotopologs considered here, N_{vib} is the same, but in principle the number could vary between different isotopologs. We fit to all three isotopologs simultaneously, using weights derived from the experimental uncertainties. We find the best fit for $N_{\text{vib}} = 77$ with a reduced chi-squared $\chi_\nu^2 = 1.3$. For $N_{\text{vib}} = 76$ and 78 we find $\chi_\nu^2 = 25$ and 26 respectively.

The final fitted parameters are given in Table 8.2, with their uncertainties and sensitivities [23]. As this is a very strongly correlated fit, rounding the fitted parameters to their uncertainties introduces very large errors in the calculated levels, so the parameters are given to a number of significant figures determined by their sensitivity [23] to allow accurate reproduction of the binding energies. The fitted value of C_6

Table 8.2 Fitted parameters from the least-squares fit and derived properties of the potential. Uncertainties are 1σ. The sensitivity is as defined in Ref. [23]

Parameter	Value	Uncertainty	Sensitivity
A/E_h	13.8866515	0.3	2×10^{-7}
$C_6/E_h a_0^6$	3463.2060	6	2×10^{-4}
$C_8/E_h a_0^8$	502560.625	7000	5×10^{-3}

is within 3% of the value from Tang's combining rule [6]. The ground-state binding energies calculated from the fitted interaction potential are included in Table 7.1.

The statistical uncertainties in the potential parameters are very small. However, our model is somewhat restrictive and the uncertainties in quantities derived from the potential are dominated by model dependence. To quantify this, we have explored a range of different models; these include using different values of β and adding an attractive exponential term in the fit to the electronic structure calculations. The estimates of uncertainties due to model dependence given below are based on the variations observed in these tests. Further measurements of more deeply bound vibrational states would be necessary to determine the details of the short-range potential.

8.3.1 Scattering Properties

Using the fitted potential, it is possible to predict scattering lengths for all isotopologs. These are given in Table 8.3. In this case the uncertainties from statistics and model dependence are comparable, though the latter are larger. The scattering lengths are also shown as a function of reduced mass in Fig. 8.9, along with both observed and calculated binding energies. The cube root of the binding energy varies almost linearly with reduced mass for an interaction potential with $-C_6/R^6$ long-range behaviour [24], except for a small curvature very near dissociation due to the Gribakin-Flambaum correction [16] of $\pi/8$ to the WKB quantization condition at threshold.

The scattering lengths are in remarkably good agreement with our previous estimates based on interspecies thermalisation in Chap. 6. The higher precision of the measurements presented here yield lower uncertainties on the scattering length and an absolute determination of the number of bound states supported by the potential.

Six of the isotope combinations have scattering lengths between $-2\bar{a}$ and $2\bar{a}$. The exception is Cs+^{176}Yb which possesses a very large scattering length due to the presence of an additional vibrational level near threshold. The moderate values of the scattering length of four of the bosonic Yb isotopes allows the production of miscible two species condensates [36] with Cs at the magnetic field required to minimize the Cs three-body loss rate [49]. Conversely, the large positive scattering length for Cs+^{176}Yb is likely to result in an enhancement of the widths of Feshbach resonances [6]. The very small interspecies scattering length of Cs+^{173}Yb indicates

Table 8.3 Interspecies scattering lengths calculated from the fitted interaction potential. Both statistical uncertainties (1σ) and estimated uncertainties from model dependence are given

Mixture	$a_{CsYb}(a_0)$	Statistical uncertainty (a_0)	Model dependence (a_0)
Cs+^{168}Yb	165.98	0.15	0.4
Cs+^{170}Yb	96.24	0.08	0.2
Cs+^{171}Yb	69.99	0.08	0.3
Cs+^{172}Yb	41.03	0.12	0.5
Cs+^{173}Yb	1.0	0.2	1.0
Cs+^{174}Yb	−74.8	0.5	3
Cs+^{176}Yb	798	7	40

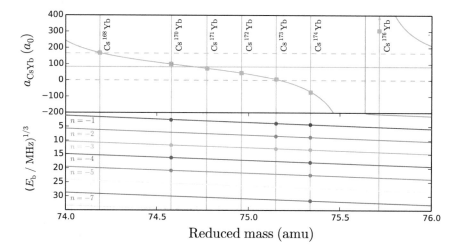

Fig. 8.9 Interspecies scattering length (upper panel) and binding energies (lower panel) for CsYb as a function of reduced mass, calculated using the fitted interaction potential. Points show measured levels; error bars are smaller than the points on this scale. The vertical lines correspond to the stable Yb isotopes. The horizontal lines on the upper figure correspond to $a = 0$, $\bar{a}$, and $2\bar{a}$

that the degenerate Bose-Fermi mixture would be essentially non-interacting. In contrast, the scattering length of $70\,a_0$ for Cs+^{171}Yb is ideal for sympathetic cooling of ^{171}Yb to degeneracy [39, 45], overcoming the problem of the small intraspecies scattering length [22] that makes direct evaporative cooling ineffective.

The miscibility of a two-species Cs+Yb condensate may be easily controlled by tuning the Cs scattering length. The most favourable isotopic combination of ^{133}Cs and ^{174}Yb may be tuned across the miscible-immiscible phase transition by a simple ramp of the magnetic field from 19 to 18 G. Allowing access to the strongly immiscible regime without significant inelastic losses that are present when tuning the inter-particle scattering length in the vicinity of a Feshbach resonance [43]. In a bichromatic trap, the interspecies overlap is controllable through the balance of bichromatic trap powers and also by the magnetic field gradient due to Yb's

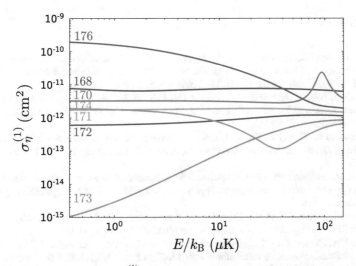

Fig. 8.10 Calculated cross sections $\sigma_\eta^{(1)}$ for interspecies thermalisation of Cs with Yb, as a function of collision energy E. The curves are labelled by the Yb isotope which is paired with ^{133}Cs

insensitivity to magnetic fields. Therefore, this system is very promising for the investigation of pattern formation dynamics in two-component condensates [25, 26]. In addition, the attractive interspecies interactions between ^{133}Cs and ^{174}Yb are extremely promising for the formation of heteronuclear quantum droplets [7, 8].

Using our fitted molecular potential it is also possible to calculate the cross sections $\sigma_\eta^{(1)}$ that characterise interspecies thermalisation [15], just like we did in Chap. 6. Figure 8.10 shows the calculated transport cross section including all relevant partial waves as a function of collision energy, for all the isotopic combinations. The low-energy cross sections vary across more than 4 orders of magnitude. Cs+^{173}Yb has a very small cross section at low energy, due to its tiny zero-energy scattering length, but this increases rapidly with energy due to both effective-range effects and p-wave scattering. A d-wave shape resonance is observed for Cs+^{170}Yb at $E/k_B =$ 90 μK. Cs+^{174}Yb has a negative scattering length at zero energy and exhibits a Ramsauer-Townsend minimum near 30 μK, where the energy-dependent scattering length crosses zero. This suggests the efficiency of sympathetic cooling for this isotopolog will be poor at temperatures above 10 μK. This will not be significant in our experiment as Cs can be cooled to temperatures well below 10 μK using DRSC but for experiments wishing to use magnetically trapped Cs this may be problematic.

References

1. van Abeelen FA, Verhaar BJ (1999) Determination of collisional properties of cold Na atoms from analysis of bound-state photoassociation and Feshbach resonance field data. Phys Rev A 59(1):578–584. https://doi.org/10.1103/physreva.59.578

2. Abraham ERI, McAlexander WI, Gerton JM, Hulet RG, Côté R, Dalgarno A (1996) Singlet s-wave scattering lengths of ^{6}Li and ^{7}Li. Phys Rev A 53:R3713–R3715. https://doi.org/10.1103/PhysRevA.53.R3713

3. Bergmann K, Theuer H, Shore B (1998) Coherent population transfer among quantum states of atoms and molecules. Rev Mod Phys 70(3):1003. https://doi.org/10.1103/RevModPhys.70.1003

4. Bohn JL, Julienne PS (1996) Semianalytic treatment of two-color photoassociation spectroscopy and control of cold atoms. Phys Rev A 54(6):R4637–R4640. https://doi.org/10.1103/physreva.54.r4637

5. Brewer RG, Hahn EL (1975) Coherent two-photon processes: transient and steady-state cases. Phys Rev A 11(5):1641–1649. https://doi.org/10.1103/physreva.11.1641

6. Brue DA, Hutson JM (2013) Prospects of forming ultracold molecules in $^2\Sigma$ states by magnetoassociation of alkali-metal atoms with Yb. Phys Rev A 87(5):052,709. https://doi.org/10.1103/physreva.87.052709

7. Cabrera CR, Tanzi L, Sanz J, Naylor B, Thomas P, Cheiney P, Tarruell L (2017) Quantum liquid droplets in a mixture of Bose-Einstein condensates. Science 359(6373):301–304. https://doi.org/10.1126/science.aao5686

8. Cheiney P, Cabrera CR, Sanz J, Naylor B, Tanzi L, Tarruell L (2018) Bright soliton to quantum droplet transition in a mixture of Bose-Einstein condensates. Phys Rev Lett 120(135):301. https://doi.org/10.1103/PhysRevLett.120.135301

9. Cohen-Tannoudji C (2015) Dark resonances from optical pumping to cold atoms and molecules. Phys Scr 90(8):088,013. https://doi.org/10.1088/0031-8949/90/8/088013

10. Debatin M, Takekoshi T, Rameshan R, Reichsöllner L, Ferlaino F, Grimm R, Vexiau R, Bouloufa N, Dulieu O, Nägerl HC (2011) Molecular spectroscopy for ground-state transfer of ultracold RbCs molecules. Phys Chem Chem Phys 13(42):18,926–18,935. https://doi.org/10.1039/C1CP21769K

11. Dutta S, Pérez-Ríos J, Elliott DS, Chen YP (2017) Two-photon photoassociation spectroscopy of an ultracold heteronuclear molecule. Phys Rev A 95(1):013,405. https://doi.org/10.1103/physreva.95.013405

12. Martinez de Escobar YN, Mickelson PG, Pellegrini P, Nagel SB, Traverso A, Yan M, Côté R, Killian TC (2008) Two-photon photoassociative spectroscopy of ultracold ^{88}Sr. Phys Rev A 78(062):708. https://doi.org/10.1103/PhysRevA.78.062708

13. Fano U (1961) Effects of configuration interaction on intensities and phase shifts. Phys Rev 124(6):1866–1878. https://doi.org/10.1103/physrev.124.1866

14. Fleischhauer M, Imamoglu A, Marangos JP (2005) Electromagnetically induced transparency: optics in coherent media. Rev Mod Phys 77(2):633–673. https://doi.org/10.1103/revmodphys.77.633

15. Frye MD, Hutson JM (2014) Collision cross sections for the thermalization of cold gases. Phys Rev A 89(052):705. https://doi.org/10.1103/PhysRevA.89.052705

16. Gribakin GF, Flambaum VV (1993) Calculation of the scattering length in atomic collisions using the semiclassical approximation. Phys Rev A 48(1):546–553. https://doi.org/10.1103/PhysRevA.48.546

17. Gunton W, Semczuk M, Dattani NS, Madison KW (2013) High-resolution photoassociation spectroscopy of the ^{6}Li$_2$$A(1^1\Sigma_u^+)$ state. Phys Rev A 88(062):510. https://doi.org/10.1103/PhysRevA.88.062510

18. Guo M, Vexiau R, Zhu B, Lu B, Bouloufa-Maafa N, Dulieu O, Wang D (2017) High resolution molecular spectroscopy for producing ultracold absolute ground-state ^{23}Na^{87}Rb molecules. Phys Rev A 96(5):052,505. https://doi.org/10.1103/PhysRevA.96.052505

19. Guttridge A, Frye MD, Yang BC, Hutson JM, Cornish SL (2018) Two-photon photoassociation spectroscopy of csyb: ground-state interaction potential and interspecies scattering lengths. Phys Rev A 98(022):707. https://doi.org/10.1103/PhysRevA.98.022707
20. Hutson JM (1993) BOUND computer program, version 5. Distributed by collaborative computational project no. 6 of the UK Engineering and Physical Sciences Research Council
21. Jones KM, Lett PD, Tiesinga E, Julienne PS (1999) Fitting line shapes in photoassociation spectroscopy of ultracold atoms: a useful approximation. Phys Rev A 61(012):501. https://doi.org/10.1103/PhysRevA.61.012501
22. Kitagawa M, Enomoto K, Kasa K, Takahashi Y, Ciuryło R, Naidon P, Julienne PS (2008) Two-color photoassociation spectroscopy of ytterbium atoms and the precise determinations of s-wave scattering lengths. Phys Rev A 77(1):012,719. https://doi.org/10.1103/physreva.77.012719
23. Le Roy RJ (1998) Uncertainty, sensitivity, convergence, and rounding in performing and reporting least-squares fits. J Mol Spectrosc 191(2):223–231. https://doi.org/10.1006/jmsp.1998.7646
24. Le Roy RJ, Bernstein RB (1970) Dissociation energy and long-range potential of diatomic molecules from vibrational spacings of higher levels. J Chem Phys 52(8):3869–3879. https://doi.org/10.1063/1.1673585
25. Lee KL, Jørgensen NB, Liu IK, Wacker L, Arlt JJ, Proukakis NP (2016) Phase separation and dynamics of two-component Bose-Einstein condensates. Phys Rev A 94(013):602. https://doi.org/10.1103/PhysRevA.94.013602
26. Liu IK, Pattinson RW, Billam TP, Gardiner SA, Cornish SL, Huang TM, Lin WW, Gou SC, Parker NG, Proukakis NP (2016) Stochastic growth dynamics and composite defects in quenched immiscible binary condensates. Phys Rev A 93(023):628. https://doi.org/10.1103/PhysRevA.93.023628
27. Lounis B, Cohen-Tannoudji C (1992) Coherent population trapping and Fano profiles. J de Phys II 2(4):579–592. https://doi.org/10.1051/jp2:1992153
28. Mark MJ, Danzl JG, Haller E, Gustavsson M, Bouloufa N, Dulieu O, Salami H, Bergeman T, Ritsch H, Hart R, Nägerl HC (2009) Dark resonances for ground-state transfer of molecular quantum gases. Appl Phys B 95(2):219–225. https://doi.org/10.1007/s00340-009-3407-1
29. Meniailava DN, Shundalau MB (2017) Multi-reference perturbation theory study on the CsYb molecule including the spin-orbit coupling. Comput Theor Chem 1111:20–26. https://doi.org/10.1016/j.comptc.2017.03.046
30. Moal S, Portier M, Kim J, Dugué J, Rapol UD, Leduc M, Cohen-Tannoudji C (2006) Accurate determination of the scattering length of metastable helium atoms using dark resonances between atoms and exotic molecules. Phys Rev Lett 96(023):203. https://doi.org/10.1103/PhysRevLett.96.023203
31. Münchow F, Bruni C, Madalinski M, Gorlitz A (2011) Two-photon photoassociation spectroscopy of heteronuclear YbRb. Phys Chem Chem Phys 13(42):18,734. https://doi.org/10.1039/c1cp21219b
32. Ni KK, Ospelkaus S, de Miranda MHG, Pe'er A, Neyenhuis B, Zirbel JJ, Kotochigova S, Julienne PS, Jin DS, Ye J (2008) A high phase-space-density gas of polar molecules. Science 322(5899):231–235. https://doi.org/10.1126/science.1163861
33. Orriols G (1979) Nonabsorption resonances by nonlinear coherent effects in a three-level system. Nuovo Cimento B 53(1):1–24. https://doi.org/10.1007/bf02739299
34. Pachomow E, Dahlke VP, Tiemann E, Riehle F, Sterr U (2017) Ground-state properties of ca_2 from narrow-line two-color photoassociation. Phys Rev A 95(043):422. https://doi.org/10.1103/PhysRevA.95.043422
35. Portier M, Leduc M, Cohen-Tannoudji C (2009) Fano profiles in two-photon photoassociation spectra. Faraday Discuss 142:415. https://doi.org/10.1039/b819470j
36. Riboli F, Modugno M (2002) Topology of the ground state of two interacting Bose-Einstein condensates. Phys Rev A 65(6):063,614. https://doi.org/10.1103/PhysRevA.65.063614
37. Rom T, Best T, Mandel O, Widera A, Greiner M, Hänsch TW, Bloch I (2004) State selective production of molecules in optical lattices. Phys Rev Lett 93(073):002. https://doi.org/10.1103/PhysRevLett.93.073002

38. Rvachov TM, Son H, Park JJ, Ebadi S, Zwierlein MW, Ketterle W, Jamison AO (2018) Two-photon spectroscopy of the ^{23}Na^{6}Li triplet ground state. Phys Chem Chem Phys 20(7):4739–4745. https://doi.org/10.1039/c7cp08481a
39. Taie S, Takasu Y, Sugawa S, Yamazaki R, Tsujimoto T, Murakami R, Takahashi Y (2010) Realization of a SU (2) × SU (6) system of fermions in a cold atomic gas. Phys Rev Lett 105(19):190,401. https://doi.org/10.1103/PhysRevLett.105.190401
40. Taie S, Watanabe S, Ichinose T, Takahashi Y (2016) Feshbach-resonance-enhanced coherent atom-molecule conversion with ultranarrow photoassociation resonance. Phys Rev Lett 116(043):202. https://doi.org/10.1103/PhysRevLett.116.043202
41. Tang KT, Toennies JP (1984) An improved simple model for the van der Waals potential based on universal damping functions for the dispersion coefficients. J Chem Phys 80(8):3726–3741. https://doi.org/10.1063/1.447150
42. Thakkar AJ, Smith VH (1974) On a representation of the long-range interatomic interaction potential. J Phys B At Mol Phys 7(10):L321. https://doi.org/10.1088/0022-3700/7/10/004
43. Tojo S, Taguchi Y, Masuyama Y, Hayashi T, Saito H, Hirano T (2010) Controlling phase separation of binary Bose-Einstein condensates via mixed-spin-channel Feshbach resonance. Phys Rev A 82(033):609. https://doi.org/10.1103/PhysRevA.82.033609
44. Tsai CC, Freeland RS, Vogels JM, Boesten HMJM, Verhaar BJ, Heinzen DJ (1997) Two-color photoassociation spectroscopy of ground state rb$_2$. Phys Rev Lett 79:1245–1248. https://doi.org/10.1103/PhysRevLett.79.1245
45. Vaidya VD, Tiamsuphat J, Rolston SL, Porto JV (2015) Degenerate Bose-Fermi mixtures of rubidium and ytterbium. Phys Rev A 92(043):604. https://doi.org/10.1103/PhysRevA.92.043604
46. Vanhaecke N, Lisdat C, T'Jampens B, Comparat D, Crubellier A, Pillet P (2004) Accurate asymptotic ground state potential curves of Cs$_2$ from two-colour photoassociation. Eur Phys J D 28(3):351–360. https://doi.org/10.1140/epjd/e2004-00001-y
47. Vitanov NV, Rangelov AA, Shore BW, Bergmann K (2017) Stimulated Raman adiabatic passage in physics, chemistry, and beyond. Rev Mod Phys 89(1):015,006. https://doi.org/10.1103/revmodphys.89.015006
48. Wang H, Nikolov AN, Ensher JR, Gould PL, Eyler EE, Stwalley WC, Burke JP, Bohn JL, Greene CH, Tiesinga E, Williams CJ, Julienne PS (2000) Ground-state scattering lengths for potassium isotopes determined by double-resonance photoassociative spectroscopy of ultracold ^{39}K. Phys Rev A 62(5):052,704. https://doi.org/10.1103/physreva.62.052704
49. Weber T, Herbig J, Mark M, Nägerl HC, Grimm R (2003) Three-body recombination at large scattering lengths in an ultracold atomic gas. Phys Rev Lett 91(123):201. https://doi.org/10.1103/PhysRevLett.91.123201
50. Winkler K, Thalhammer G, Theis M, Ritsch H, Grimm R, Denschlag JH (2005) Atom-molecule dark states in a Bose-Einstein condensate. Phys Rev Lett 95(6):063,202. https://doi.org/10.1103/PhysRevLett.95.063202
51. Zanon-Willette T, de Clercq E, Arimondo E (2011) Ultrahigh-resolution spectroscopy with atomic or molecular dark resonances: exact steady-state line shapes and asymptotic profiles in the adiabatic pulsed regime. Phys Rev A 84(6):062,502. https://doi.org/10.1103/physreva.84.062502
52. Zuchowski PS, Hutson JM (2010) Reactions of ultracold alkali-metal dimers. Phys Rev A 81(6):060,703. https://doi.org/10.1103/PhysRevA.81.060703

Chapter 9
Conclusions and Outlook

9.1 Summary

In this thesis I have reported the first measurements of the binding energies of ground state CsYb molecules and the scattering lengths of the Cs+Yb system. This represents a major milestone in the experiment and is essential for devising the most efficient route to creation of rovibrational ground state CsYb molecules.

These experiments were performed on an ultracold optically trapped mixture of Cs and Yb atoms. The setup of the optical dipole trap and the schemes for efficient loading of Yb isotopes into the optical trap have been described. Evaporative cooling of bosonic ^{174}Yb resulted in the reliable production of large condensates containing $3 - 4 \times 10^5$ atoms. The changes required for cooling fermionic ^{173}Yb were discussed and the production of a six-component degenerate Fermi gas of 8×10^4 ^{173}Yb atoms with a temperature of $0.3\,T_F$ was reported. The production of a ^{174}Yb BEC and a ^{173}Yb DFG are the first observations in the UK of quantum degeneracy in a closed-shell atom.

The implementation of effective cooling methods for Cs was also discussed. Degenerate Raman sideband cooling was implemented which allowed a large number of Cs atoms to be cooled to below $2\,\mu$K and polarised efficiently in the $|F = 3, m_F = +3\rangle$ state. To enable efficient transfer into the same optical trap as Yb, we described the setup of a large volume 'reservoir' trap. By evaporatively cooling Cs in the dimple trap we produced a Cs BEC containing 5×10^4 atoms. We have demonstrated the tunability of the Cs BEC by using a low-field Feshbach resonance to control the BEC expansion during TOF.

Further, we have described the loading of Cs and Yb atoms into the same optical dipole trap and presented the first measurements of the scattering properties of Cs+Yb. A kinetic model is derived to describe the observed thermalisation and is used to extract elastic scattering cross sections. Comparison of the thermalisation cross sections with quantum scattering calculations allowed us to evaluate scattering lengths for all seven CsYb isotopologs.

We have characterised the production of ultracold electronically excited Cs_2 and CsYb molecules using one-photon photoassociation in an optical dipole trap. The setup of a laser system with a large tunable frequency range for photoassociation

© Springer Nature Switzerland AG 2019
A. Guttridge, *Photoassociation of Ultracold CsYb Molecules
and Determination of Interspecies Scattering Lengths*, Springer Theses,
https://doi.org/10.1007/978-3-030-21201-8_9

is described and we have demonstrated control over the Cs_2 photoassociation rate by tuning the Cs intraspecies scattering length. We have directly observed the distance dependence of the hyperfine coupling constant in electronically excited CsYb molecules and precisely measured the binding energies of vibrational levels in the 2(1/2) potential for multiple CsYb isotopologs. The analysis of these binding energy measurements using an extended Le Roy-Bernstein model allowed the number of bound states supported by the 2(1/2) potential to be determined to be 154 or 155.

Finally, we have described the production of vibrationally-excited ground state CsYb molecules using two-photon photoassociation. Using the two-photon photoassociation technique we have measured the binding energies of high-lying vibrational levels of $^{133}Cs^{174}Yb$, $^{133}Cs^{173}Yb$ and $^{133}Cs^{170}Yb$ molecules including the last-bound state. Autler-Townes spectroscopy of these bound-bound transitions provided the first experimental results on the transition strengths, important for the identification of two-photon routes to the ground state. We have reported the fitting of a ground-state interaction potential based on electronic structure calculations to the measured binding energies. The model potential is used to calculate the scattering lengths of all CsYb isotopes to much greater precision than the thermalisation measurements. The model potential may also be used to predict positions of novel interspecies Feshbach resonances. Magnetoassociation using these predicted Feshbach resonances followed by STIRAP is a promising route to the creation of ground state paramagnetic CsYb molecules.

9.2   Outlook

9.2.1   Short Term

Double Degeneracy in Cs-Yb

The measurements reported in this thesis inform our choice of pathway to the ground state. The two-photon measurements yield essential information on both the positions of Feshbach resonances and which atomic mixtures have favourable properties for cooling to degeneracy. The measured interspecies scattering lengths are favourable for sympathetic cooling of Cs using most Yb isotopes. The exceptions being ^{176}Yb which has a very large interspecies scattering length and ^{173}Yb which has a very small interspecies scattering length. As we have seen throughout this thesis, both ^{133}Cs and multiple Yb isotopes are capable of being independently cooled to degeneracy. Therefore, using elastic collisions between the species it should be possible to enhance the number of degenerate atoms in comparison to cooling the species independently. Increasing the number of distinct mixtures which are capable of being cooled in the setup gives a diversified approach to the production of CsYb molecules. In addition, a binary condensate comprised of Cs and the most abundant Yb isotope, ^{174}Yb, is promising for the investigation of the miscible-immiscible phase transition in a heteronuclear system and has the potential to form mass-imbalanced quantum droplets.

The major obstacle to the achievement of efficient sympathetic cooling in the current setup is the large difference in polarisabilities of the two species at 1070 nm. This effect can be mitigated by the installation of a bichromatic trap using 532 nm light, in addition to the 1070 nm light already in use. The 532 nm light leads to an attractive potential for Yb ($\alpha_{Yb} = 261\,a_0^3$) and a repulsive potential for Cs ($\alpha_{Cs} = -224\,a_0^3$). Therefore, the relative difference in trap depth can be tuned by changing the relative powers of 1070 and 532 nm trapping beams. This trapping arrangement also allows the compensation of the differential gravitational sag that may hamper attempts of sympathetic cooling.

During the writing of this thesis the 532 nm trap has been installed in the setup. A rendering of the science chamber with the bichromatic trap beams is shown in Fig. 9.1 along with the calculated trapping potentials for Cs and Yb. The trap can be tailored depending on the mixture under consideration. For example, we can obtain large samples of ^{174}Yb atoms, therefore, it may be beneficial to have a slightly larger trap depth for Cs to allow cooling of Cs by sympathetic evaporation of ^{174}Yb. Conversely, a larger trap depth for Yb would be beneficial for a different mixture with contrasting scattering properties, such as the fermionic Cs+^{171}Yb mixture where the Yb isotope has an extremely small intraspecies scattering length but the mixture has a moderate interspecies scattering length. The larger trap depth for Yb would allow ^{171}Yb to be sympathetically cooled by Cs. Further, with the implementation of the Cs reservoir

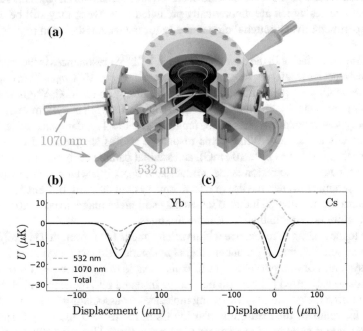

Fig. 9.1 Implementation of a Bichromatic trap. **a** 3D rendering of the science chamber with bichromatic trapping beams. Lower plot: Contributions of the 532 nm (green) and 1070 nm (orange) beams to the total trap depth (black) of **b** Yb and **c** Cs trapping potentials. The intensities used in the simulation were $I_{532} = 22.3$ kW/cm^2 and $I_{1070} = 11.7$ kW/cm^2

trap we have large sample of $N > 1 \times 10^7$ Cs atoms at $T \sim 3\,\mu$K, a large portion of which are lost during the dimple loading. It may be efficient to use this large sample of cold atoms to perform initial sympathetic cooling of Yb down to the temperature of the reservoir atoms, regardless if subsequent evaporation is performed using Yb or Cs.

In summary, there are a large number of sympathetic cooling regimes waiting to be explored and many are promising for the production of double-degenerate mixtures of Cs and Yb. The significant enhancement of the mixture's PSD and overlap in the bichromatic trap will be beneficial in the search for novel interspecies Feshbach resonances.

Magnetoassociation of CsYb

It has been well established by the work on bi-alkali systems that magnetoassociation and STIRAP is the most favourable method for creation of rovibrational ground state molecules deep in the ultracold regime. The CsYb interaction potential gained from fitting to the two-photon photoassociation measurements predicts multiple interspecies Feshbach resonances at low magnetic fields ($B < 200$ G) for multiple isotopes. The widths of these low field resonances are predicted to be fairly narrow ($\Delta < 1$ mG) which may limit their observability. However, we note that in the RbSr experiment features with theoretical widths of 70 μG were observed for Mechanism II and 90 nG features were observed for Mechanism III [2]. This suggests these resonances which are theoretically predicted to be weak may still be observable experimentally, although their suitability for magnetoassociation remains to be explored.

An important factor in the observation of the ^{87}Rb^{87}Sr resonances is the extremely large interspecies scattering length of the mixture $a_{87,87} > 1600\,a_0$, which enhances the observability of the features. This may be troublesome in CsYb as the most promising resonances are associated with the ^{133}Cs$+^{173}$Yb system, where the extremely low interspecies scattering length $a_{133,174} = 1\,a_0$ dampens our expectations for this isotopolog. While some resonance widths are predicted to be fairly large for these systems ($\Delta \simeq 100$ mG), an essential parameter for assessing the efficiency of magnetoassociation is the product $\Delta \times a_{\mathrm{CsYb}}$ [8], which is not favourable for this system. However, the use of a fermionic system may still be beneficial as the Pauli blocking of collision induced losses may still make magnetoassociation in this mixture more favourable than in bosonic mixtures.

Due to the scaling of resonance width with magnetic field strength it is likely that a set of bias coils capable of producing higher magnetic fields will need to be installed in the system. For this purpose a set of coils capable of producing ~ 1800 G have been wound and tested [9] but still need to be installed into the experiment. We are also exploring the possibility of forming molecules by associating Cs in another spin state. This removes the restriction of using Cs in the $|F = 3, m_F = +3\rangle$ state and enables a larger number of resonances to be considered. However, this imposes an additional constraint that the inelastic two-body loss rate must be low in the region of magnetic field around the predicted interspecies resonance. This is challenging in the particular case of Cs, as the two-body inelastic loss rate is high at most values of the

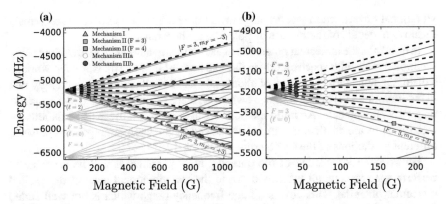

Fig. 9.2 a Level crossing diagram for ^{133}Cs^{173}Yb. Binding energy from centre of weight of the hyperfine structure versus magnetic field strength. The dashed black lines are atomic levels, the solid lines are molecular levels, with cyan corresponding to the $\ell = 0$ levels of the lower ($F = 3$) hyperfine manifold, red the $\ell = 2$ levels of this manifold and green corresponding to the upper ($F = 4$) hyperfine manifold ($\ell = 0$ levels). The atomic level of $|F = 3, m_F = +3\rangle$ is highlighted in blue. Markers denote level crossings that produce resonances, the relevant coupling is indicated by the marker shape: triangles I, squares II and circles III (see Sect. 2.3). **b** Enhanced view of level crossing diagram for the fields currently achievable in the experiment

magnetic field. However, with the assistance of Jeremy Hutson and his group we have identified some interspecies resonances at higher field where the two-body inelastic loss rate is low, some of which are in the easily accessible $|F = 3, m_F = +2\rangle$ state.

In the initial search for Feshbach resonances it is likely we will search in a mixture involving ^{173}Yb due to our demonstrated ability to cool this isotope to degeneracy and the larger number of resonances at low magnetic fields. The predicted Feshbach resonances for the Cs^{173}Yb isotopolog are plotted in Fig. 9.2. The blue dashed line shows the atomic energy of the $|F = 3, m_F = +3\rangle$. The multiple resonances that arise for this state are labelled by a marker representing the coupling mechanism responsible. It is evident that there are multiple lower field resonances that should be observable with the coil set currently installed in the experiment.

9.2.2 Long Term

STIRAP of associated Feshbach molecules requires the identification of an intermediate state with strong coupling to both initial and final states. The photoassociation measurements presented earlier form a basis for theoretical models of ground and excited molecular potentials to allow identification of a suitable intermediate state. However, a model of the molecular potential beyond the primarily long-range potential considered here is required to accurately predict the Franck-Condon overlap at small interatomic distances. This is because of the highly oscillatory behaviour of the vibrational wavefunction at short range makes the Franck-Condon overlap extremely sensitive to details of the molecular potential at short range.

In bi-alkalis the molecular potentials of excited states have been well studied using traditional molecular spectroscopy in thermal beams [1, 3, 6, 10]. However, work on a diatomic molecules composed of alkali and closed-shell atoms is exceedingly rare in the literature [12] and before our own work there was no published spectroscopy on the CsYb molecule. Ab initio methods are also not much use, as calculations for these systems are challenging, this is evidenced by the large discrepancies between different predictions of the CsYb ground state [5, 13, 14, 17].

For accurate knowledge of the potential at short range, it will be essential to perform spectroscopy all the way down to the last vibrational level of the excited molecular potential. This covers a huge frequency range which is not well suited for study with an optically trapped atomic mixture due to the long cycle time of the experiment, however, this has been performed recently in the NaLi system [15]. Measurements of this nature are more suited to traditional molecular spectroscopy such as Fourier-transform infrared spectroscopy (FTIR) in a heatpipe. These experiments are more challenging with mixtures of alkali and closed-shell atoms due to their distinct atomic properties. The major complication in spectroscopy of these molecules is the large background signal from dimers formed from the alkali atom.

Recent work involving spectroscopy of RbSr molecules on helium nanodroplets [12] is encouraging for performing extensive spectroscopy on alkali-closed-shell systems. Spectroscopy of higher electronic states which approach the Yb triplet manifold at long range may guide future experiments on alternative optical association techniques for CsYb molecules.

Taking a more long-term view, the installation of an optical lattice will be essential for the investigation of lattice spin models with ground state CsYb molecules and will be required to protect the molecules from decay due to exchange reactions. The lattice will enable efficient preparation of molecules using magnetoassociation and will allow the use of Cs atoms in any internal state due to the protection from intraspecies collisions. In addition, two degenerate gases confined in a lattice is a promising platform for studying novel atomic physics in lattices [4, 7, 11, 16].

9.3 Concluding Remarks

The measurements of the Cs+Yb scattering lengths presented in this thesis allows us to devise the optimal scheme for producing a high PSD mixture of Cs and Yb in a lattice and enables the prediction of interspecies Feshbach resonances for magnetoassociation. Creation of quantum degenerate mixtures of Cs and Yb using the techniques laid out in this thesis are favourable for a large number of Cs+Yb isotopic mixtures, many of which show promise for realising interesting Bose-Bose and Bose-Fermi mixtures with tunable interactions. This is encouraging for loading an optical lattice of preformed Cs+Yb pairs that will allow efficient association of molecules. In addition to the favourable scattering properties of the mixture, the prospects for magnetoassociation are also promising, with many more Feshbach resonances accessible than originally predicted at the outset of the project. The next steps in the experiment promise to be very exciting and should consolidate CsYb as a great candidate for the production of paramagnetic molecules in alkali-closed-shell systems.

References

1. Alps K, Kruzins A, Tamanis M, Ferber R, Pazyuk EA, Stolyarov AV (2016) Fourier-transform spectroscopy and deperturbation analysis of the spin-orbit coupled $a^1\Sigma_+$ and $b^3\Pi$ states of KRb. J Chem Phys 144(14):144,310. https://doi.org/10.1063/1.4945721
2. Barbé V, Ciamei A, Pasquiou B, Reichsöllner L, Schreck F, Zuchowski PS, Hutson JM (2018) Observation of Feshbach resonances between alkali and closed-shell atoms. Nat Phys 14:881–884. https://doi.org/10.1038/s41567-018-0169-x
3. Birzniece I, Nikolayeva O, Tamanis M, Ferber R (2012) B(1)$^1\Pi$ state of KCs: high-resolution spectroscopy and description of low-lying energy levels. J Chem Phys 136(6):064,304. https://doi.org/10.1063/1.3683218
4. Bruderer M, Klein A, Clark SR, Jaksch D (2008) Transport of strong-coupling polarons in optical lattices. New J Phys 10(3):033,015. https://doi.org/10.1088/1367-2630/10/3/033015
5. Brue DA, Hutson JM (2013) Prospects of forming ultracold molecules in $^2\Sigma$ states by magnetoassociation of alkali-metal atoms with Yb. Phys Rev A 87(5):052,709. https://doi.org/10.1103/physreva.87.052709
6. Docenko O, Tamanis M, Ferber R, Pazyuk EA, Zaitsevskii A, Stolyarov AV, Pashov A, Knöckel H, Tiemann E (2007) Deperturbation treatment of the $a^1\Sigma^+$ - $b^3\Pi$ complex of NaRb and prospects for ultracold molecule formation in $X^1\Sigma^+(v=0; j=0)$. Phys Rev A 75(042):503. https://doi.org/10.1103/PhysRevA.75.042503
7. Griessner A, Daley AJ, Clark SR, Jaksch D, Zoller P (2006) Dark-state cooling of atoms by superfluid immersion. Phys Rev Lett 97(22):220,403. https://doi.org/10.1103/PhysRevLett.97.220403
8. Hodby E, Thompson ST, Regal CA, Greiner M, Wilson AC, Jin DS, Cornell EA, Wieman CE (2005) Production efficiency of ultracold Feshbach molecules in bosonic and fermionic systems. Phys Rev Lett 94(120):402. https://doi.org/10.1103/PhysRevLett.94.120402
9. Kemp SL (2017) Laser cooling and optical trapping of ytterbium. PhD thesis, Durham University
10. Kim JT, Kim B, Stwalley WC (2014) Analysis of the alkali metal diatomic spectra. Morgan & Claypool Publishers. https://doi.org/10.1088/978-1-6270-5678-6
11. Klein A, Bruderer M, Clark SR, Jaksch D (2007) Dynamics, dephasing and clustering of impurity atoms in Bose-Einstein condensates. New J Phys 9(11):411. https://doi.org/10.1088/1367-2630/9/11/411
12. Lackner F, Krois G, Buchsteiner T, Pototschnig JV, Ernst WE (2014) Helium-droplet-assisted preparation of cold RbSr molecules. Phys Rev Lett 113(153):001. https://doi.org/10.1103/PhysRevLett.113.153001
13. Meniailava DN, Shundalau MB (2017) Multi-reference perturbation theory study on the CsYb molecule including the spin-orbit coupling. Comput Theor Chem 1111:20–26. https://doi.org/10.1016/j.comptc.2017.03.046
14. Meyer ER, Bohn JL (2009) Electron electric-dipole-moment searches based on alkali-metal- or alkaline-earth-metal-bearing molecules. Phys Rev A 80(4):042,508. https://doi.org/10.1103/physreva.80.042508
15. Rvachov TM, Son H, Park JJ, Notz PM, Wang TT, Zwierlein MW, Ketterle W, Jamison AO (2018) Photoassociation of ultracold ^{23}Na ^{6}Li. Phys Chem Chem Phys 20(7):4746–4751. https://doi.org/10.1039/c7cp08480c

16. Schmidt R, Knap M, Ivanov DA, You JS, Cetina M, Demler E (2018) Universal many-body response of heavy impurities coupled to a Fermi sea: a review of recent progress. Rep Prog Phys 81(2):024,401. https://doi.org/10.1088/1361-6633/aa9593
17. Shao Q, Deng L, Xing X, Gou D, Kuang X, Li H (2017) Ground state properties of the polar alkali-metal-ytterbium and alkaline-earth-metal-ytterbium molecules: a comparative study. J Phys Chem A 121(10):2187–2193. https://doi.org/10.1021/acs.jpca.6b11741

Printed in the United States
By Bookmasters